T0185936

Universities as Engines of Economic Development

Edward Crawley • John Hegarty
Kristina Edström • Juan Cristobal Garcia Sanchez

Universities as Engines of Economic Development

Making Knowledge Exchange Work

 Springer

Edward Crawley
Massachusetts Institute of Technology
Cambridge, MA, USA

John Hegarty
Trinity College Dublin
Dublin, Ireland

Kristina Edström
KTH Royal Institute of Technology
Stockholm, Sweden

Juan Cristobal Garcia Sanchez
Massachusetts Institute of Technology
Cambridge, MA, USA

ISBN 978-3-030-47551-2   ISBN 978-3-030-47549-9   (eBook)
https://doi.org/10.1007/978-3-030-47549-9

This Springer imprint is published by the registered company Springer Nature Switzerland AG
The registered company address is: Gewerbestrasse 11, 6330 Cham, Switzerland

To Karen, Neasa, Adam, Ana Paula, and all the members of our families who supported us in this endeavor.

Authors' Foreword

We wrote this book because we identified an opportunity to enhance the sustainable development of society. We seek to raise universities' ambitions and capabilities in a way that is aligned with the desires of society. Advancing societal development, and, in particular, economic development, is not a new contribution of universities. They already do this through education, research, and what is traditionally called service. We saw a need for systematic exploration of these matters across universities worldwide. Our contribution is captured as a set of patterns of human behavior that we call *practices*. We also draw attention to the interplay between these practices and illustrate them with case studies.

These ideas are born out of joint reflection on various practical experiences, harvesting some of our own lessons learned.

Edward Crawley has been involved in founding and reforming university efforts since 1993. When he served as the Co-Director of the Cambridge (UK)—MIT Institute, he observed the potential of all universities to improve and that they would benefit by using a framework to organize and communicate their efforts. He applied and evolved the framework during his service as Founding President of the Skolkovo Institute of Science and Technology.

John Hegarty was president for ten years of Trinity College, Dublin, a comprehensive university with strong culture and traditions. He led a major expansion of research, growth in interdisciplinary and inter-institutional collaboration, a transformation of organizational structures, and engagement with external stakeholders. Since then, he has applied his knowledge by helping other universities and organizations. He is deeply committed to the importance of all university disciplines, and of a balance between curiosity-driven and use-inspired research.

Kristina Edström approached these issues through 20 years of experience in developing educational programs and faculty competence at KTH Royal Institute of Technology, as well as in national and international collaborations and consultancies. Her motivation is to promote meaningfulness in education, to empower students for doing something with their understanding, and thus preparing them for a stimulating and productive working life with positive impact on society.

Juan Cristobal Garcia Sanchez has conducted innovation programs for universities, high-tech companies, and industrial clusters. He has written or edited 20 books. He develops educational approaches for innovation that prepare change agents who can build learning organizations, and who excel in systemic thinking, creative collaboration, and agile execution. He strives for a sustainable and inclusive society by strengthening the links between learning, entrepreneurship, and prosperity.

This book is intended to be useful and pragmatic, highlighting opportunities for addressing practical issues of economic development. The readers we have in mind are not primarily specialist scholars, and this text will not pass the tests of rigorous social science research. Still, we hope that researchers of university development can appreciate a practice-based and action-oriented reflection.

Primarily, we write for those interested in a comprehensive discussion about practical approaches. Our framework can serve as a point of departure for those who need to make decisions and take action. Broadly speaking, the ideas provide an agenda for deep engagement with stakeholders and useful guidance that allows agreement on an effective way forward. The intended readers are those who have a stake in universities:

- University leaders—board members, administrators, and faculty thought leaders. They will see a resource and reference point in planning university contributions to economic development.
- Funders—government, industry, philanthropies, and alumni. They will have a clearer understanding of how the resources of the university translate into impact on economic development.
- Educational policy makers—governments, and quality and accreditation organizations. They will better understand how universities will evolve and the latitude they need.
- Economic development agencies—They will better understand how universities will contribute to economic development.
- Partners—industry, small and medium enterprise, and government organizations. They will better understand how to engage with the university for mutual benefit.
- Neighbors—communities, local government, business, and the media. They will better understand how the university can engage with them to strengthen the local community and its development.
- Students and their families—They will learn how to contribute to the evolution of the university and how to benefit from a deeper involvement in knowledge exchange.

The book has as its foundation a set of patterns of human behavior that we call practices. There are many dimensions to the activities within a university. To make this book practical, it was important to describe a set of implementable practices within a consistent framework.

By illustrating the practices with actual cases, we demonstrate that the practices are real and representative. The practices illustrate what many universities around the world are doing. We could have chosen from among hundreds of other cases at many other excellent institutions. Based on our own insights and experiences, we selected 43 cases representing effective deployment. Each is adapted to the local context, conditions, and culture. Some of the cases are well documented and internationally recognized, while others are little known. They show a diversity in institutions and are widely distributed geographically. The scope varies from institution-wide activities to work on a smaller scale. The cases are a substantial and valuable part of this book. We are greatly indebted to those who provided the examples that demonstrate feasibility and help give life to the narrative.

We acknowledge that today is a time of accelerating change for universities. This book will help all readers understand the current state of practice, based on a retrospective journey. If we were now to start the 20 years of work that this book summarizes, we would, for example, have a stronger emphasis on sustainable development in all its dimensions.

This book focuses on economic contributions based on science, engineering, and entrepreneurship, by technical universities and comprehensive universities that have technology and science programs. This reflects the majority of our own direct experience, and it is also the area where the current expectations are the greatest. We do not in any way seek to devalue the other forms of contributions by universities, or the contributions by other types of higher educational institutions.

Foreword

Universities are among the oldest institutions in the world. Many are certainly older than most governments or businesses. It is sometimes said that they have lasted so long because they are built upon long-lasting codes of behavior and well-tested traditions. The same reasons are put forward to explain why some academics resist change. Indeed, there are general principles of academic freedom, intellectual rigor, and integrity that have stood for centuries, even in contentious times. In my view, these principles alone do not explain the remarkable longevity of some universities.

Successful universities have lasted so long because they have been prepared continually to adapt to changing circumstances. Change is generally supported by students, who are mostly young and insist on their teachers staying up to date, but change is sometimes resisted by academics although they frequently come up with ideas for change themselves in their research. Change has also been needed to cope with the ten times increase in young people seeking degrees over the last 50 years.

This book points out that it is in the nature of successful universities to adapt to changes in society, and the authors use a wide range of case studies to show how this has been achieved. The authors have used their extensive international experience to pull together a reference volume that will be valuable for anyone running or creating universities in the twenty-first century.

Another questionable assumption that gained credibility in the past is that academics preserve their creativity by avoiding worldly influences, especially those related to commerce and money. In fact, changes related to commerce and money do penetrate into universities and have done so at an accelerating pace over the last few decades. These changes have particularly affected research in the Science, Technology, Engineering, and Mathematics (STEM) disciplines, but have also occurred in many areas of the humanities and social sciences.

STEM research has been affected because the intense competition created by globalization has forced almost all industrial and business organizations to abandon basic research. The bulk of the resources they have made available for research and development are now needed to keep up with the development of modern highly complex products and technologies. Fundamental research on subjects unrelated to their products, such as that pursued at the AT&T and IBM research laboratories in the middle of the twentieth century, is rarely found in industry today. Even basic research underpinning products and technologies is increasingly conducted in partnerships with universities, where it can gain financial support from government and other non-company sources.

To encourage and support these partnerships, universities have had to acquire new equipment and research staff to be relevant to industry and business. Money for this has come from governments, charities, alumni, and, in a few cases, from profits made on teaching. To justify the money, universities have had to explain how their research benefits society. In some institutions, this met initially with strident resistance. The requirement for research to have an "impact" was scorned. Fortunately, on further consideration, universities realized that most of their research did have an impact, and the extra funds allowed them to expand into exciting new areas of research.

The authors explain that universities need not undergo revolutionary change to positively impact the societies of which they are part. Many are already strengthening their contributions to sustainable economic, social, and cultural developments. In other words, they have discovered that they can act as engines of economic development when they accelerate innovation in industry and enterprise through knowledge exchange.

While universities often prioritize knowledge transfer, some still think of it as a one-way flow of mostly intellectual property. Few have made knowledge exchange one of their core activities as defined in this book. The authors point out that activities that strengthen knowledge exchange can be built into education, research, and catalyzing innovation without requiring radical action. It can be accomplished simply by further evolution of the university's traditional strengths.

Knowledge exchange is not an add-on or separate function, but can be embedded in the university's core education, research, and innovation missions. It implies softening the borders of the university and lowering barriers to make the flow of people, discoveries, and creations easier. This proves of benefit to everyone. Knowledge exchange taken deeply into the thinking and actions of the university can improve its own intellectual capabilities as well as accelerating innovation in its partners.

The novel contribution of this book is that it charts a systematic and pragmatic approach to strengthen knowledge exchange with well-defined effective practices and a large number of case studies to illustrate how this has been achieved by a broad range of universities across the world. These case studies guide those intent upon increasing the impact of their efforts.

Universities will be relieved that the authors are not saying radical change is necessary. But they are challenging university leaders to rethink the common patterns of behavior in education, research, and innovation. They show how the support functions of the university need to reposition themselves to better support knowledge exchange using the same pragmatic approach. It signals to all partners of the university what they should do to play their part in the exchange.

The book delves into the sensitive area of how universities can measure their progress towards successful knowledge exchange, and how goals and individual behavior can be reconciled.

There is a pragmatic and broad-ranging discussion of the detailed mechanisms that can be used to achieve change in existing universities. The authors point out that this requires an understanding of how traditional collegial governance, especially in older universities, has weakened as universities have grown and fragmented into smaller communities in different units and locations. They emphasize that it is nonetheless important to involve individual academics in the process of deciding upon, and making, these changes.

The authors consider national university systems and the need to encourage diversity of mission, pointing out that optimum intellectual contribution does not flow from undifferentiated universities. This is a danger that can result from formulaic schemes of funding. There should instead be mechanisms that allow universities to excel in different fields of endeavor and collect around them the contemporary thought leaders and their students. Competition among institutions should be encouraged.

Finally, they emphasize that continuous interaction with stakeholders is essential in founding new universities, as is the choice of the first head of the university and the team of senior academics that support the leader.

This is not a volume that needs to be read from beginning to end although this is well worthwhile for anyone interested in the progression of universities. At the end of the first chapter is an outline of what is covered in each of the following chapters. Each chapter concludes with a valuable summary of what has been discussed within the chapter. This allows the reader to rapidly explore the book for ideas and methods for achieving the sort of change they are seeking. Most established universities are already in the process of adapting to changing circumstances and will have their own experiences to compare with those presented in the case studies. As the authors point out, they must also consider their own unique cultures and the needs of the societies they serve.

To summarize, this book is a comprehensive reference volume that explores how universities can maximize their contributions to societal change. It is a timely contribution that pulls together international experience in keeping universities relevant to the societies they live in, while they continue to fulfill their responsibilities to educate and to seek truth through scholarship and research.

University of Cambridge, February 2020   Alec Broers

Alec Broers, the Lord Broers, FRS, FMedSci, FREng, is a member of the UK House of Lords and was Chairman of its Science and Technology Committee. Previously, he served as the President of the Royal Academy of Engineering and as the Vice Chancellor of the University of Cambridge.

Preface and Summary

Synopsis: How Can Universities Better Contribute to Sustainable Development?

When universities participate in systematic knowledge exchange, they can become more effective, powerful engines of economic development. Knowledge exchange, the give-and-take of people, capacities, and ideas between the university and its partners accelerates innovation in industry and enterprise, leading to tangible, and often profound, economic impact. Universities can re-envision and embrace knowledge exchange as an intrinsic part of the core academic activities: education, research, and catalyzing innovation. This re-envisioning does not require disruptive action. It can be accomplished by building on universities' traditional strengths and boldly quickening the pace of the institutions' evolution.

Expectations for University Engagement Are High

Society faces challenges, from small to grand. Governments, industry, and civil society all strive to achieve sustainable economic, social, and cultural development, for a growing population—all while protecting the environment.

Universities affect people's lives in broad and significant ways. They are understood to be engines for society's sustainable development and a nation's competitiveness. Their contributions are recognized, and significant public and private funds are being invested in increasing these contributions. Universities' external stakeholders seek to more productively engage universities in addressing the needs of society, particularly in economic development. Expectations are high.

Universities Can more Effectively Contribute to Economic Development

University thought leaders are aware of stakeholder expectations, and many are attempting to strengthen university engagement. We believe that if it chooses to, an ambitious university can significantly increase its impact on society.

There is widespread interest on how science and technology can contribute more directly to economic development: by providing more knowledgeable and skilled graduates, making discoveries in research that could underpin new technologies and high value jobs and explicitly engaging in catalyzing innovation.

Our book focuses on how research universities, through their science, technology, and related activities, can strengthen their impact on economic development. Our observations and recommendations are intended to help universities and their stakeholders.

This focus does not reduce the importance of other disciplines to economic development, or the importance of all disciplines to the social and cultural well-being of society. We leave it to others to explore these relevant contributions. Our focus is intended to make this book's scope bounded and its proposals understandable and implementable.

Knowledge Exchange Has the Central Role in this Strengthened Engagement

Knowledge exchange is the to-and-fro movement of people, capabilities, and ideas across the university's porous boundary. It is the process of engagement between a university and its external partners. Crossing boundaries is critical to effective knowledge exchange.

From the point of view of economic development, the primary partners are industry, small and medium enterprise, and government. Industry and enterprise take on knowledge from the university, hire its graduates, and create and markets products, services, and systems. Government creates the policy framework for higher education institutions, funds university activities in the public interest, and benefits from university outcomes in its various public bodies and institutions.

Knowledge Exchange Accelerates Innovation, Contributing to Economic Development

Our fundamental proposition is that universities are more effective as engines of economic development when they participate in systematic knowledge exchange. The exchange of knowledge accelerates innovation in industry, small and medium enterprise and government organizations, leading to new goods, services, and systems. Their introduction creates real economic impact. If we wish to strengthen the economic impact of universities, we must re-envision knowledge exchange. The reader is encouraged to read Chapter 1 for details on the impact of universities discussed so far.

Knowledge Exchange Gives an Expanded Role to Academic Activities

The Fig. 1 indicates the core academic activities in education, research, and catalyzing innovation that take on an expanded role. These activities craft the outcomes that flow to knowledge exchange:

- *Talented graduates*—knowledgeable and capable students and citizens whose preparation is the outcome of education
- *Discoveries*—the new facts, data, theories, and models that are the outcomes of research
- *Creations*—the outcomes of catalyzing innovation: new artifacts, concepts, prototypes, inventions, know-how, and business methods

Strengthening knowledge exchange in the university is not about establishing a new unit or organization. Rather, it requires reimagining the patterns of behavior, or practices, in education, research, and catalyzing innovation. These practices are built on the traditional mission and values of the university. It does not subvert tradition; it lets the university concentrate on what it does best, leaving to its partners to do what they do best.

Our approach to strengthening knowledge exchange is pragmatic. Our recommendations are largely derived from the real experience of the authors and are consistent with the scholarship in higher education. We offer specific proposals for action—broadly applicable, but not prescriptive. They are references that a university can consider and adapt to its context and aspirations.

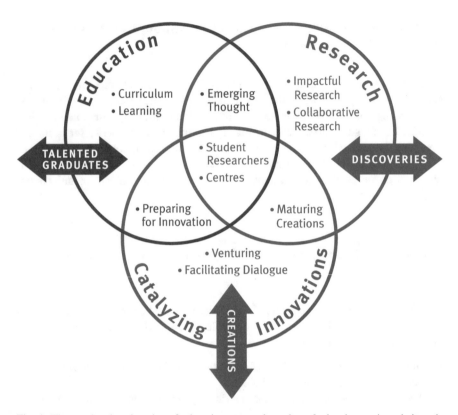

Fig. 1 The overlapping domains of education, research, and catalyzing innovation, their main outcomes (graduates, discoveries, and creations), and the 11 academic practices

Systematic, Effective, and Adaptable Practices Build Capability in Knowledge Exchange

We advocate a *systematic approach* to knowledge exchange by considering an integrated and cross-disciplinary set of academic practices, each with a means of stakeholder engagement and proactive exchange of outcomes.

As shown in the Fig. 1, we identify 11 *effective academic practices* in the overlapping domains of education, research, and catalyzing innovation. Each is supported by a rationale and key actions for implementation. The academic practices are effective when they yield outcomes that amplify knowledge exchange.

These academic practices are enabled by supporting practices at the university and its partners which we call a *framework for the adaptable university*. This includes six supporting practices that facilitate the effective operation of the university, as well as approaches to program evaluation, alignment with industry, government, and philanthropies, and to leading change. The readers are invited to read Chapter 2 for more details on the systematic approach.

Forty-three Case Studies of Effective Practice Provide Important Examples

For each of the academic and supporting practices, we identify case studies that are representative of successful deployment by universities across the world—43 in total. We learned from these cases and offer them as examples of effective practice. We choose technical and comprehensive research universities, old and new, large and small, and with a wide geographic spread. It is a sampling of many different university systems and cultures that share an aspiration for greater economic contribution.

Three of the cases present longer explanations of how the practices are smoothly integrated into a coherent whole. These highlight the accomplishments of the Singapore University of Technology and Design, University College London, and Pontificia Universidad Católica de Chile. The other 40 cases highlight the adaptation of individual practices.

Education Contributes to Knowledge Exchange

We now outline the 11 academic practices that span education, research, and catalyzing innovation. These form a systematic and integrated approach to strengthening knowledge exchange, as suggested by the Fig. 1.

In education, the following practices prepare students to be *talented graduates*, acting as agents of knowledge exchange, as innovators and as citizens:

- Implementing an integrated curriculum, preparing students in disciplinary fundamentals, and essential life and professional skills
- Engaging students in active, experiential, and digital learning for deeper conceptual understanding, self-efficacy, and self-learning
- Expeditiously introducing emerging and cross-disciplinary thought from research into the curriculum as new disciplines or interdisciplinary programs
- Offering students courses within the curriculum in leadership, management, and entrepreneurship to better prepare them for innovation

Research Contributes to Knowledge Exchange

Research leads to *discoveries*, often revealing phenomena or truths that previously existed but were unknown or ill explained. These discoveries have the potential to impact knowledge exchange and innovation through:

- Pursuing fundamental discoveries along a spectrum, from curiosity-driven to use-inspired, that impact both scholarship and society
- Collaborating within and across disciplines in search for new high-impact cross-disciplinary discoveries and fields of thought

- Empowering large-scale Centres of Research, Education, and Innovation to find directly implementable solutions to the pressing societal issues
- Energizing research by engaging undergraduate and postgraduate students, preparing them as agents of knowledge exchange

Catalyzing Innovation Contributes to Knowledge Exchange

In catalyzing innovation, participants produce *creations*. These can be synthesized objects, processes, and systems that have never before existed. Increased impact on knowledge exchange occurs by:

- Maturing the technical readiness of discoveries and creations within the university and assessing their business readiness
- Facilitating dialogue and formal agreements with partners to promote the adoption of discoveries and creations
- Engaging in the actual entrepreneurship process within the university to create new ventures and better prepare entrepreneurs

The reader is invited to read Chapters 3–6 for an in-depth discussion of the core academic practices of education, research, and catalyzing innovation.

Adaptable Universities Provide Support for the Academic Practices

We have identified six supporting practices that characterize a university desiring to strengthen knowledge exchange. They make up part of the framework for an adaptable university:

- Engaging external stakeholders to understand their needs and to inform the curriculum, research, and innovation agendas
- Evolving the university culture to be supportive of activities leading to economic development
- Revising the university's mission, strategy, and priorities to focus investment of resources and to communicate how the university will distinguish itself including in innovation
- Updating governance procedures to strengthen the role in knowledge exchange and innovation
- Recruiting and developing faculty and staff who will strengthen knowledge exchange and engage in the innovation mission
- Ensuring that academic facilities are functionally suitable to new learning, innovation, and collaborative research activities

Adaptable Universities Evaluate Progress and Set Faculty Expectations

How can universities evaluate their progress as they embark on a revised path? Evaluation is a necessity, given the high level of interest in outcomes. Evaluation is well embedded in some universities and systems, but not uniformly. In addition, setting expectations for the contributions of individual faculty will help align the faculty actions with the university goals.

We identify two practices which add to the adaptable framework:

- Program Evaluation—Collecting evidence that reflects the university goals, and evaluating the success of programs and units, demonstrating the contributions of the university
- Faculty Expectations and Recognition—Setting expectations and recognizing accomplishments of individuals in education, research, innovation, and knowledge exchange

Partners Align their Practices to Support the University

To be effective, the actions of the university will need complementary actions by its main partners: industry and enterprise, government, and philanthropies and alumni. We advocate three partner practices leading to better alignment, also part of the adaptive framework:

- Understanding the university's needs and capabilities
- Building up the university's capacity to contribute
- Developing partners' capacity to co-create and absorb outcomes from the university: talented graduates, research discoveries, and innovation creation

Change is Possible and Necessary

The need to adapt and adopt these practices implies that ongoing change is necessary. For many universities, this will require considerable change. Even institutions already well positioned will need some effort. This change and adaptation need not be disruptive and is within the capability of universities that have a long history of evolution and adaptation. The wrinkle in the twenty-first century is that the pace of change quickening.

Change Process is Adapted for Existing Universities, Systems, and New Universities

Existing universities possess strongly embedded cultures. These will influence the dynamics of their change. Universities need to base their plans for change on an understanding of important strengths in their university culture. These include collegiality, thought leadership, evidence, benchmarking, piloting, and academic rhythms in time.

For university systems interested in adapting these approaches to a network of institutions, it is worthwhile considering what constitutes a strong system: ambition greater than the sum of the parts, complementarity of missions, differentiation to allow excellence, competition to avoid complacency, and collaboration to achieve economies of scale.

At new universities that are just being founded, everything has to be created including the culture itself. This requires a careful founding process, definition of vision, staged implementation, and enthusiastic champions.

The topics on supporting practices, evaluation, alignment, and change are presented in Chapters 7–10.

An Invitation

We invite the reader to first scan the book, gaining an understanding of the systematic approach to knowledge exchange, the academic practices and case studies, and the framework for an adaptable university. Following this pre-read, a thorough read can start from the beginning, or concentrate on the reader's topics of interest. In this regard, the book is designed to be modular.

We expect the book will be both useful and unsettling. Useful in that it supports views likely already developed by the reader, but not yet acted upon. Unsettling in that it stretches the university, advancing the boundaries of its potential contributions. Change is essential, but not always comfortable.

Manchester, MA   Edward Crawley
Dublin, Ireland   John Hegarty
Stockholm, Sweden   Kristina Edström
Belmont, MA   Juan Cristobal Garcia Sanchez

Acknowledgements

We gratefully acknowledge the help and advice of more than one-hundred colleagues and friends who engaged with us in this project and selflessly participated in drafting and reviewing this book.

Key contributors who helped shape our ideas and frameworks included Ruth Graham, Peter Gray, Mike Gregory, Mats Hanson, Michael Kelly, William Lucas, Johan Malmqvist, Mikhail Myagkov, Mats Nordlund, and Graeme Reid.

Other colleagues who brought unique perspectives and ideas included Duane Boning, Howard Califano, Claude Canizares, Tony Chan, Michael Cima, Charles Cooney, Mike Crow, Jose Estabil, Clement Fortin, Ellen Hazelkorn, Alan Hughes, Thomas McCarthy, Lesley Millar-Nicholson, Edward Roberts, Bruce Tidor, and Lap-Chee Tsui.

In addition, many thoughtful reviewers supported us through their critical and constructive input: Patrick Aebischer, Phillip Altbach, Barry Bozeman, Robert Braun, John Jianzhong Cha, Glyn Davis, James Duderstadt, Henry Etzkowitz, Gerhard Fasol, Billy Fredriksson, Daniel Frey, Isak Frumin, Malcolm Grant, Akira Ishibashi, David Lloyd, Thomas Magnanti, Eric Mazur, Thomas McCarthy, Richard Miller, Mille Milnert, Fiona Murray, Seán Ó Foghlú, Richard Roth, Warren Seering, John Van Maanen, and Frans van Vught.

We thank the numerous colleagues who generously assisted as case authors and contributors: Pernille Andersson, Jorge Baier, M. Balakrishnan, Robert Braun, Jane Butler, Tom Byers, Michael Cardew-Hall, Alan Chan, Chong Tow Chong, Michael Crow, Juan Carlos de la Llera, Barry Dwolatzky, Simon Edström, Mikael Enelund, Yves Gnanou, Constance Fleet, Ruth Graham, Cory Hallum, Matthew Harvey, Isabel Hilliger, Alison Holmes, Erik Hultén, Anna-Karin Högfeldt, Steven Imrich, Wayne Kaplan, Michael Kelly, Anthony Kenyon, Anette Kolmos, Benjamin Koo, Karl Koster, Johan Malmqvist, Rob Martello, Román Martínez Martínez, Eric Mazur, John Mitchell, Emily Moore, Justin Lee Mynar, Mats Nordlund, Diarmuid O'Brien, Nick Oliver, Sebastian Pfotenhauer, Nina Qvistgaard, Jochen Runde, Ambuj Sagar, Rona Samler, Sanjay Sarma, Sung-Chul Shin, Shuli Shwartz, Lynn

Andrea Stein, Toshihiro Tanaka, Toshitsugu Tanaka, Claus Thorp Hansen, Ernst Ludwig Winnacker, and Dick K.P. Yue. Cases were also created by colleagues at Mediacom, EPFL, and the Hong Kong University of Science and Technology.

Contents

Chapter 1
The Impact of Universities on Economic Development

1.1 Introduction and Overview of our Approach

1.1.1 Expectations Are High

Universities have significant and broad impact on society and receive substantial investments of public and private funds. External stakeholders seek to better engage universities in addressing the needs of society, particularly in economic development. These stakeholders understand that some universities are moving the economic needle. Universities are central to the policies of regions and nations wishing to boost their economic, social, and cultural development, and do it in a sustainable way. Expectations are high.

University thought leaders know of these stakeholder expectations, and many are attempting to strengthen university engagement with society. They are more carefully considering the needs of society in the evolution of the university. These thought leaders would benefit from tangible resources indicating how universities can continue to adapt to meet stakeholder needs.

1.1.2 Our Contributions Can Help Universities to Adapt

We believe that, if it chooses to, a university can increase its impact on society in general, and on economic development in particular. Doing so *does not require a disruptive departure from tradition, but a reaffirmation and extension of the long-standing missions of the university.* Our contributions here are crafted to assist universities and their stakeholders in this effort. We offer a systematic, pragmatic, and broad-based approach.

Our scope covers the wide range of university activities, including education, research, and catalyzing innovation, as shown in Box 1.1. However, our treatment is

© Springer Nature Switzerland AG 2020
E. Crawley et al., *Universities as Engines of Economic Development*,
https://doi.org/10.1007/978-3-030-47549-9_1

focused on how science and technology engage with economic development at research-led universities that embrace innovation. Many of our findings broadly apply to other types of institutions, other disciplines, and all spheres of societal development.

Our fundamental proposition is that universities are most effective as engines of economic development when they accelerate innovation in industry, small and medium enterprise, and government organizations through *systematic exchange of knowledge*. Knowledge exchange is the to-and-fro exchange of people, capacities, and ideas across the porous boundary of the university. Crossing boundaries is critical to effective knowledge exchange.

We identify a set of effective *academic practices that foster knowledge exchange*. These academic practices expand the traditional role of education, research and catalyzing innovation to emphasize knowledge exchange. The practices are built on overlapping patterns of behavior in education, research and catalyzing innovation. The practices produce outcomes that cross the boundaries of the university. We illustrate these practices with 43 cases from universities and institutions around the world.

Some change at universities will be needed to strengthen their engagement with society. Our perspective is that universities have long demonstrated an ability to adapt, but at rates that were relatively slow. To accelerate this rate of change, we develop *a framework for the adaptable university*. This framework consists of supporting practices (e.g., strategy, culture) as well as approaches to evaluation, partner alignment, and leading change. Assuming a shared vision and distributed leadership, this framework can usefully support the evolution of the university.

Box 1.1 The Core Domains of Activity of a Modern University

Universities sponsor activity by faculty and students in three overlapping core academic domains, each of which can yield new knowledge:
- *Education* is the process of increasing the knowledge and skills of students, (e.g., in physics, medicine, or entrepreneurship courses).
- *Research* is the process of discovery at the frontiers of knowledge, and the quest for an increased understanding of aspects of our world previously unknown or imperfectly explained (e.g., through a laboratory or social science experiment).
- *Catalyzing Innovation* encompasses the activities of the university around creativity and synthesis, which lead to creations that have never existed (e.g., a piece of music, a sketch of a new building, an invention).

We use the term "catalyzing innovation" to denote creative activities *within* the university and to distinguish them from the more general sense of innovation as the translation of ideas into tangible artifacts, products, services, and systems that occur broadly in society.

Sometimes universities use the term *service* to describe their mission on behalf of society. But *service* also refers to the faculty's important contributions to university governance and scholarly communities. For our purposes, the term *engagement* is more useful than *service*. It denotes the mutually beneficial interaction of the university with its external stakeholders in education, research, and catalyzing innovation.

1.1.3 Our Approach Is Pragmatic

Our approach to addressing society's expectations for universities is pragmatic. We intend this book to be a resource, a reference point, and a set of proposals for consideration by ambitious universities and their stakeholders. The literature on universities is already rich, and generally known to university leaders, policy makers, and stakeholders. This book is different. It is about change. It is aimed at university leaders who feel responsible to convert aspirations to practice, to government policy makers who need to implement wise policy, and to external stakeholders who need to figure out the practicalities of working with universities.

We suggest positive actionable proposals. These proposals are broadly applicable and not prescriptive. They invite consideration. They can be adapted to the context and aspirations of any specific university.

Our work is firmly based in the scholarship on higher education, some of which is acknowledged in Box 1.2. These references have shaped our work and constitute a good reading list. Hundreds of other citations follow. But our observations and recommendations are largely derived from our personal experiences (Box 1.3).

Our proposals are set on the intermediate level, somewhere between a high-level discussion of university policy and a detailed analysis of a university operation. We build on the long-established values and strengths of the university, focusing on what universities traditionally have done best.

Box 1.2 A Brief Overview of some of the Important Literature on Universities

The published literature has much that is useful. One important body of work comes from scholars of higher education. These include: Altbach and Salmi on research universities [1], Clark on the characteristics of entrepreneurial universities [2], and Goddard, Hazelkorn et al. on reclaiming the concept of a university deeply engaged in its community [3].

Useful contributions also come from past and present university leaders. These include Rhodes on the challenges to the US research university and the need for new thinking [4], Duderstadt on the arrival of the knowledge economy and the need for change [5], and Crowe on the design and development of a new university model for the US system [6]. These writers put forward high-level visions and discuss interaction with policy.

The Organization for Economic Cooperation and Development (OECD) has published many reports on the role of the university in a knowledge economy, especially related to innovation and commercialization of research [7]. It has also developed a guiding framework for entrepreneurial universities [8].

A set of published works from the Glion Colloquium has examined issues for the research university in depth and covered similarities and differences between the USA and Europe [9–12]. These provide useful comparative frameworks and important data.

Box 1.3 The Basis of our Observations and Recommendations

Our observations and recommendations are derived from:
- The rich literature on higher education.
- The personal experience of the authors, who have founded a university, led two universities, collectively held appointments at ten, and advised dozens of others.
- An extensive set of case studies developed for this book that illustrate the broad applicability of the proposed practices.
- Benchmarking studies that identify global leading practices and trends [13, 14].
- The experience of the Cambridge—MIT institute [15], which accepted the explicit mission of understanding and codifying approaches to accelerating innovation in the UK.
- The international Conceive Design Implement Operate (CDIO) Initiative [16], which developed standards for engineering education now used in more than 100 universities worldwide.

MIT's institution-building experiences, most recently in Portugal [17], Abu Dhabi [18], and Singapore [19].

1.1.4 The Main Contribution Is an Actionable Agenda

The main contribution of this book is to *identify and codify a set of activities, illustrated with 43 case studies, that collectively exemplify how universities can exchange knowledge with partners more effectively. This would accelerate innovation and entrepreneurship and contribute to sustainable economic development.* More specifically, we describe:

1. *A systematic approach to knowledge exchange* between the university and its partners. Partners are stakeholders with whom the university is actively exchanging knowledge. The approach includes: identifying the partners and their needs; developing a sensitivity to these needs in the conduct of education, research, and catalyzing innovation; and a proactive process to exchange the knowledge outcomes (Chapter 2).
2. Concrete *effective academic practices* in education, research, and catalyzing innovation, each of which is accompanied by a rationale and key actions for implementation. The academic practices yield outcomes that amplify knowledge exchange (Chapters 3–5). The *integration* of the practices in a program is the driver of knowledge exchange (Chapter 6).
3. *A framework for the adaptable university.* This includes six enabling practices that support the effective operation of the university (stakeholder engagement,

mission, governance, culture, staff, and facilities), as well as approaches to program evaluation, to alignment with industry, government, and philanthropies, and to leading change (Chapters 7–10).

Throughout this book, we provide 43 inspirational cases that demonstrate the applicability of these practices. These are drawn from a wide range of universities, large and small, technical and comprehensive, new and old, and from every inhabited continent. Of the cases, 40 largely illustrate specific practices, activities, and principles. Three more comprehensive cases demonstrate the integration of education, research, and catalyzing innovation.

To see how this agenda fits within the historical context, in Section 1.2 we turn to the history of universities as resilient and adaptable institutions, followed by a summary of their economic contributions in Section 1.3. Section 1.4 holds the main assertion of the book: that universities' economic impact arises from knowledge exchange. In Section 1.5, this idea is broadened to societal impact.

1.2   The Adaptability of Universities and the New Expectations

1.2.1   Based on their History of Resilience and Adaptability, Universities Are up to the Challenge

The evolution of universities is an extraordinary tale of small beginnings, constant organic growth, resilience, and adaptation to sometimes rapidly evolving societal regimes. The longevity of universities is likely due to this resilience and adaptability. This history of adaptability prepares the university to address the expectations of today's society [20].

Throughout the world, ancient cultures founded academic institutions that were often affiliated with the state or religious entities. What we now think of as independent universities in the European model began in the eleventh century in Bologna [21]. The primary mission of these institutions was the *preservation and distribution of knowledge* [22]. In today's parlance, these teaching-led institutions provided for the *democratization of knowledge* [23].

A second mission, *developing new knowledge,* evolved over time from ancient roots, particularly in Alexandria. In the early nineteenth century, Humboldt proposed a university free from any interference, yet state guaranteed [24]. As a central element, it included the unfettered pursuit of new knowledge through research. Many European universities developed according to this ideal. In the USA, educators combined the undergraduate teaching college with the graduate school into a new model: the research university [25].

In the twentieth century, a third mission developed among the research universities—that of *applying knowledge* [26]. This emerged from desires to work closer with partners in industry, government, and international organizations on problems of commercial and societal importance. The objective was to transfer knowledge through *application* [27].

In the third decade of the twenty-first century, the world's universities are increasingly numerous and varied. More than 20,000 universities across the world now have widely varying student populations (from hundreds to hundreds of thousands), missions, research intensities, and innovation agendas. They have many stakeholders—students, faculty, government, industry and small and medium enterprise, philanthropic donors, and, ultimately, the public. Each stakeholder comes with its own set of expectations. Each university reflects its own history and context.

Many new expectations are arising. Universities, which have evolved so much over a thousand years, are being asked to address these new expectations, and quickly. Universities must draw on their deepest traditions of resilience and adaptability to do so.

1.2.2   Governments Have Clear Expectations for the Impact of Universities on Society

The impact of universities on their regional, national, and international communities is broad, complex, and far-reaching. This impact is now central to the policies of regions and nations wishing for sustainable economic, social, and cultural development.

Governments, representing the people, have, perhaps, the broadest view. They fund higher education because it means social progress for individuals and it contributes to the common good. Universities support stable societies and contribute to competitive economies by providing talented graduates and new knowledge. Governments are particularly interested in social mobility and seek to make higher education accessible to most of their citizens. This generally requires a diversity of institutions to serve students of different interests, motivations, ages, and economic backgrounds.

Government officials—for example, legislators, heads of bureaus—are keenly interested in the impact that new knowledge can have on all aspects of society. They seek stronger engagement of universities with society. They want universities to pay closer attention to society's needs, to become more involved, and to better contribute to solutions. They believe that if universities engage with the users of knowledge, the outcomes will be more valuable goods, services, and systems, as well as stable and rewarding jobs. Nations have a strong interest in the quality of graduating students. Some have published broad qualifications frameworks, and even detailed expected learning outcomes to ensure adequate preparation of graduates.

The United Nations has its own expectations for universities. Its Millennium Development Goals of 2000 [28], and its more recent 2030 Agenda for Sustainable Development [29] include 17 Sustainable Development Goals. Together, these are intended to harmonize economic growth, social inclusion, and environmental protection. These, and other UN documents on societal development, refer to fostering innovation, creativity, entrepreneurship, and quality education. All of these have clear links to the universities' roles. The UN's statement that "without technology and innovation, industrialization will not happen, and without industrialization, development will not happen [30]" is a clear call for university participation along with participation by industry, government, and society. In this book, when we discuss development, we mean sustainable development aligned with these UN goals.

1.2.3 Universities Are Responding to these Expectations

Universities are responding to the expectations of government officials and citizens by taking steps to strengthen their engagement with society.

The impact of universities on society is broad and multifaceted and spans economic, social, and cultural development (Fig. 1.1). Influence by stakeholder engagement, university outcomes result from the hard work of the faculty and staff, students, and administrators, all making use of facilities and funding. Their work leads to a set of outcomes that play a pivotal role in knowledge exchange. The main outcomes are:

- Knowledgeable and skilled university graduates and informed citizens, whose preparation is the outcome of education. We call these *talented graduates.*
- Discoveries—the new facts, data, and theories that are the outcomes of research.
- Creations—the outcomes of catalyzing innovation at the university. They include artifacts, concepts, prototypes, inventions, methods, and know-how, as well as artistic and architectural outcomes, business methods, and medical procedures.

Universities impact society when these outcomes are exchanged with partners such as industry, small and medium enterprise, government organizations, cultural institutions, nonprofits, and other higher education institutions. These partners take actions based on university outcomes that address societal good.

Universities have less-tangible impacts as well. They are vibrant organizations whose personnel contribute to the public, commercial, and government discourse that sets the direction of society. Universities attract thinkers and doers who provide

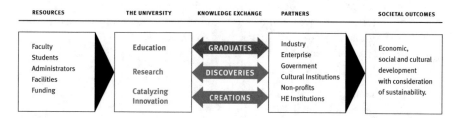

Fig. 1.1 Education, research, and catalyzing innovation, the university activities that eventually lead to societal benefit, emphasizing the central role of knowledge exchange

a rich human resource to the region. They can be good neighbors in towns and cities, enriching local life. When universities' personnel are highly diverse, that fact alone can increase empathy and tolerance between communities and improve the operation of democracy.

In general, what makes universities special is low barriers to the development and spread of new ideas. New ideas are strong agents of change for society, building social cohesion and stimulating discussions of public matters.

Universities' impact is broad and intense. To avoid mere generalizations, however, we plan to keep a tight focus here on university contributions to sustainable economic development. Certainly, universities are important contributors to social and cultural development, but we leave it to others to discuss those threads.

1.3 The Impact of Universities on Economic Development

1.3.1 The Indications of Economic Development Suggest Significant Impact

Governments and other stakeholders grasp the significant economic impact of universities. The indications show strong, consistent trends, highlighted in Box 1.4 and detailed below. Some of these reflect on economic influence through innovation, and others indicate the broader economic influence of universities.

Box 1.4 Summary of the Indications of Universities' Economic Impact

There is no single indication of the positive impact of universities on economic development, but evidence from various directions suggest substantial impact:

- The support by economic theory for government investment in universities, and for the macroeconomic value of knowledge.
- The policy recognition by global leaders who identify the path from rational investment in universities to economic return.
- The regional innovation impact that develops around universities in the form of techno-parks and new science cities.
- The actions taken by governments to strengthen the economic engagement of existing universities and to create new universities.
- The historical economic contributions by universities to regional economic growth and to local welfare.

The economic contributions by the graduates of universities, as measured by the revenues of companies they found and by jobs created.

Ever since Adam Smith, economic theory has held that government should invest in activities with long-term return for the public good [31]. Among these investments are universities, which provide public good in various forms, including economic development [32]. The link between knowledge and economic growth was highlighted in 2018 when the Nobel Prize in economics was awarded to New York University professor of economics Paul Romer for his integration of technological innovations into macroeconomic analysis. Evidence of the economic benefit of universities is documented in the USA, where most economists agree that about half of American economic growth after 1945 came from technological innovation that mostly originated in universities [33].

1.3.2 The Economic Impact of Innovation

World leaders understand the special role of innovation and its economic impact and are making appropriate policy. In 2014, Xi Jinping said "China is a nation committed to innovation-driven development. … We must prioritize innovation education [34]." In 2018, the European Commission stated: "Research and innovation (R&I) are crucial for sustaining Europe's socio-economic model and values, as well as its global competitiveness. ... Sustainable growth in the future can only come from investing in R&I now [35]."

People may debate the economic benefit of direct spending by governments on universities, but the benefits of co-investment by industry and enterprise in proximity to major research universities are obvious. Companies understand that proximity boosts collaboration [36, 37].

One need only drive down Route 101 in Silicon Valley, California to sense the enormous economic development that has occurred there in the last several decades [38]. Similar developments are found near many major research universities [39]. In some cases, it is clear that a university's initiative was key. In the 1950s, Frederick E. Terman, Stanford University's former dean of engineering and provost, actively championed the Stanford Industrial Park, which became one of the nucleators of Silicon Valley [40]. Likewise, in the 1990s, MIT President Charles M. Vest encouraged Novartis to relocate its international research center to Cambridge, Massachusetts [41]. In both cases, intense innovation-related commercial development now surrounds the campus.

Governments around the world now understand how universities contribute to economic development, and they are spending taxpayer money to foster such contributions. Local funds, equal to billions of US dollars, have been spent on just a handful of these projects in recent years. The founding of new universities in Singapore has made it the learning hub of the region. In Russia, the goals of the Skoltech and the Skolkovo Innovation Center are similar, linking science with industry and enterprise. Beijing's Tsinghua University Science Park has brought some of the world's top IT companies to its front door. At various scales, nations around the world are investing with the expectation of economic return [42]. Leaders know that having a research university in a community contributes powerfully to creating an innovation-based economy [43].

1.3.3 The Broader Economic Impact of Universities

Universities have broad regional economic impact as well. A recent comprehensive study analyzed more than 15,000 universities and found that university growth has an association with ensuing regional gross domestic product growth [44]. These conclusions coincided with studies that recognize the important role of universities in regional economic development and the formation of clusters [45]. Evidence suggests that universities have important long-term effects on the growth of their local economies [46]. In short, universities are strongly linked with regional prosperity [47].

Historically, evidence shows that places where the US government gave support for the establishment of colleges more than a century ago have more educated workforces to this day; therefore, people earn higher wages and economic activity increases [48]. Universities located in small communities also can improve local welfare [49], including supporting the economies of small towns [50]. Even small colleges with predominantly commuter populations have important effects on the communities where they are located [51].

Strikingly powerful economic effects stem from the actions of university graduates. According to studies (mostly done in North America), companies founded by university graduates tend to be economic powerhouses.

The most striking measurement of core-mission impact was carried out by Roberts and Eesley for MIT using one parameter—the scale of existing companies founded by living alumni [52]. They conservatively estimated that the contribution to the US GDP approached a staggering 10%. This contribution, if recognized as a nation, would place the "MIT Nation" as comparable in size to the entire economy of Russia or India [53]. From roots in Stanford, almost 40,000 active companies have grown up; they have created an estimated 5.4 million jobs and generate annual world revenues of $2.7 trillion [54]. Harvard alumni have founded more than 146,000 companies and organizations, operating in more than 150 countries. These enterprises account for 20.4 million jobs worldwide and have generated nearly $3.9 trillion in revenues [55].

1.4   The Economic Impact of Universities Arises from Knowledge Exchange

1.4.1   Innovation and Entrepreneurship Lead to Economic Development

How is economic development linked to the activities of the university? It happens through a chain of events (Fig. 1.2). At the highest level, people and their governments are interested in a strong economy where individuals, industry and enterprise prosper, economic opportunities expand, and public services can grow.

Sustainable development occurs when there is "prosperity and economic opportunity, and social well-being" while at the same time providing for the "protection of the environment [56]." In the terms of the Brundtland Report: "Sustainable development is development that meets the needs of the present without compromising the ability of future generations to meet their own needs [57]."

Economic development largely has its origins in innovation and entrepreneurship. The macroeconomic indicators of an economy include *economic growth* and industrial *competitiveness* [58]. Economic growth is the increase in the market value of goods and services produced by an economy. Competitiveness measures the ability to sell goods and services in a given market, compared to other goods in the same market [59].

Both growth and competitiveness depend on making and selling better goods, services, and systems, and, therefore, rely on a robust ecosystem of *innovation and entrepreneurship* (Fig. 1.2). Actual *innovation* is more than invention [60]. It involves the transfer of lower value economic resources to higher value ones through the development of goods and services. Innovation brings many benefits. It permits

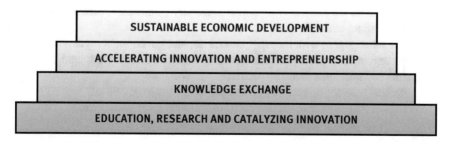

Fig. 1.2 Hierarchy of outcomes on the path from the actions of universities to economic development

rapid response to new technology and market opportunities, including the building of new companies that become employers. In this context, *entrepreneurship* can be thought of as new venture formation concurrent with the development of a first good or service.

1.4.2   Knowledge Exchange Accelerates Innovation

The main role of universities in supporting economic development is in *accelerating innovation and entrepreneurship* [61]. This suggests rethinking how new ideas are generated in the university [62] and moved swiftly through the innovation cycle from design to sales and operation.

Knowledge exchange is the key process [63] (Fig. 1.2). It is a multidirectional flow of people, capabilities, and information among relevant contributors [64]. In the case of economic development, the relevant flow is between universities and partners in industry [65], small and medium enterprise [66], and government organizations [67]. In one direction, partners identify their challenges and needs. Then, they engage in a dialog with scholars to help shape university efforts in education, research, and catalyzing innovation [68].

In the other direction, the important flow from the university is the stream of *talented graduates*, who are knowledgeable and empowered. Also flowing from the university are *discoveries* from research [69], and *creations*—our term for the outcomes of catalyzing innovation within the university. In its mission to accelerate innovation, a university's personnel must also engage with government regulators, public interest groups, investors, and others [70]. That flow of outgoing conversations is also important.

Our fundamental point is that *economic development and innovation occur as a result of effective knowledge exchange between universities and their partners*.

This knowledge exchange:

- Emerges from the main *activities at a university*—education, research, catalyzing innovation, and their interplay.
- Occurs by engaging with industry, enterprise, and government organizations.
- Gives universities a deeper understanding of stakeholder needs and the constraints and problems faced by partners.
- Stimulates relevant research discoveries and innovative creations.
- Is enabled by graduating students who carry knowledge with them.

By engaging in knowledge exchange, universities will accelerate innovation and entrepreneurship, and eventually impact economic development. In Chapter 2, we identify a set of effective practices that greatly facilitate this knowledge exchange.

1.5 The Broader Impact of Higher Education on Societal Development

1.5.1 Economic Development Linked to Research Universities

As we discuss above, the focus of this book is on economic development. We concentrate on the sustainable economic development that arises from innovations and entrepreneurship, facilitated by knowledge exchange among research universities that embrace innovation and their partners.

While all disciplines at a university can contribute to economic development [71], we focus primarily on the strong linkage between science and technology, and economic development. By *science and technology* we mean science, engineering, mathematics, and computing, and closely related aspects of the management of technology and innovation.

Many kinds of academic institutions can contribute to economic development [72], including research-led universities, teaching-led universities, research institutes, and innovation hubs [73]. In this book, we concentrate on research-led universities based on science and technology (S&T), and comprehensive universities with research programs in S&T, which, in some senses, have a broader set of outcomes to offer. We examine primarily universities of this type that *embrace innovation* and generate and exchange knowledge useful to innovation and entrepreneurship.

Though the scope of our work is S&T-based programs at research universities, our ideas will also apply to the many institutions that aspire to develop as economic engines. These ideas are also relevant to governments and foundations seeking to establish new universities of this type.

This choice of our scope does not in any way reduce the importance of contributions by universities to societal development outside the economic sphere. Other disciplines and other types of academic institutions also contribute. We leave it to others to explore these important contributions in detail.

1.5.2 All Types of Academic Institutions Foster Economic Development

Society is well served by variety in its academic institutions. For example, the economy may need high-level specialists who know how to explore new solutions, technologists who can synthesize and manage innovation, and technicians who can make things work. No one type of institution could likely provide for all of these needs.

Box 1.5 suggests that all combinations of education, research, and innovation can be found in the ecosystem of academic institution. Each of these types of institutions can play a role in accelerating innovation, and all could benefit from aspects of the approaches suggested in this book. Any type of institution can adapt the systematic approach to knowledge exchange, effective practices, and the framework for an adaptable university, as described in the following chapters.

Box 1.5 Variety in Academic Institutions Supports Knowledge Exchange

Some institutions have missions primarily in education, others in research, and still others in catalyzing innovation. Each has an important role to play in support of economic development.

- Teaching-led universities can contribute to innovation and entrepreneurship by producing many skilled graduates.
- Research institutes, such as the Max Plank Institutes and their international and corporate counterparts, including some academies of science, play an important role in making discoveries.
- Translational research-to-innovation establishments—sometimes called accelerators and catalysts, such as the Fraunhofer Institutes in Germany or the Catapult centers in the UK—help to transfer creations to the market.

Pair-wise combinations are common and important: education + research, the traditional Humboldt university; education + innovation, for example, the Singapore polytechnics; and research + innovation, a combination common in government laboratories. Sometimes these pairings exist within an institution, and sometimes through alliances. For example, a regional teaching university can cooperate with a nearby national research laboratory.

What we call *"a research university that embraces innovation"* will demonstrate the full three-way integration: education + research + innovation.

1.5.3   Societal Development Benefits from all University Disciplines

While we explain the importance of knowledge exchange through the economic lens, one can argue that knowledge exchange matters equally in all spheres of societal development. All the disciplines at a university can contribute, as suggested by Box 1.6. The pairing of science and technology with economic development is only one segment of the overall impact of universities.

Since other disciplines have broad societal impact, they may benefit from the approaches described in the following chapters. In the remainder of the book, we develop a systematic approach to knowledge exchange for economic development and describe effective practices and the framework of the adaptable university. But we are mindful that these approaches have potential for broader application.

Box 1.6 Disciplinary Impact on Economic, Social, and Cultural Development

Disciplines at a university contribute to the sustainable development of society in rich and complex ways:

- Science and engineering directly contribute to economic development through technology-based products, services, and systems. This segment is the topic of our book. But science and engineering also have influence on social well-being and on culture (e.g., through contributions to health care and the development of social media).
- Business, economics, and law contribute directly to economic development but also more widely to stable employment and social justice.
- Other disciplines in the social sciences, such as sociology, psychology, and political science help to inform evidence-based social policy. They also contribute to the economy.
- The humanities contribute to the understanding of self and others, language and communication, and to the creative industries such as literature, film, and music.
- Medicine primarily supports social well-being through health care, but as a sector it is an economic engine as well.
- Architecture and urban planning build spaces and cities for society, also contributing to art and culture.

All disciplines can impact the service sectors, including finance and banking, media, retail, and tourism.

1.6 Summary and an Overview of the Book

Worldwide, expectations are growing that universities should engage more deeply in societal development, particularly in sustainable economic development. Universities acknowledge this expectation and are moving to respond. They are resilient institutions and have demonstrated throughout their history that they can adapt.

Universities already contribute significantly to economic development. They amplify the intensity of successful innovation and entrepreneurship in the economy. Innovation is fed in part by knowledge exchange between universities and partners. The main contribution of this book is to provide resources, reference points and proposals that, if adapted locally, will strengthen knowledge exchange, innovation, and economic development.

We intend this book to be pragmatic and useful. It begins in this chapter by connecting knowledge exchange with economic development. The book continues in Chapter 2 with a proposal for a systematic approach to knowledge exchange by communication across boundaries. At the next level, eleven effective academic practices are identified and are loosely grouped into those that are primarily linked to education (Chapter 3), to research (Chapter 4), and to catalyzing innovation (Chapter 5). In the case studies, we give dozens of interesting examples of implementation of these practices around the world, and three examples of the integration of the practices (Chapter 6).

The remainder of the book develops a framework for thinking about the university as an adaptable institution. We explain a set of practices that are instrumental in the success of such adaptable institutions. These include: engaging stakeholders; evolving culture, mission, and strategic planning; improving governance; developing faculty and staff capabilities; and building academic facilities. These are presented in Chapter 7. Chapter 8 discusses the importance of, and challenges of program evaluation. In Chapter 9, we discuss the potential for alignment of universities, industry and enterprise, government and philanthropies. Chapter 10 addresses the implications for change at universities and in national systems, and the creation of new universities.

References

1. Altbach PG (2011) The past, present, and future of the research university. In: Altbach PG, Salmi J (eds) The road to academic excellence. World Bank Publications, pp 11–32
2. Clark B (1998) Creating entrepreneurial universities: organizational pathways of transformation. Emerald Group Pub Ltd
3. Goddard J, Hazelkorn E, Kempton L, Vallance P (2016) The civic university: the policy and leadership challenges. Edward Elgar Pub, Cheltenham, UK
4. Rhodes F (2001) The creation of the future: the role of the American University. Cornell University Press, Ithaca, NY
5. Duderstadt J (2009) A university for the 21st century. University of Michigan Press, Ann Arbor, MI
6. Crow M, Dabars W (2015) Designing the new American University. Johns Hopkins University Press, Baltimore, MA
7. The Organisation for Economic Co-operation and Development (2015) The innovation imperative: contributing to productivity, growth and Well-being. Organization for Economic Cooperation and Development, Paris
8. The Organisation for Economic Co-operation and Development (2012) A Guiding Framework for Entrepreneurial Universities. https://www.oecd.org/site/cfecpr/EC-OECD Entrepreneurial Universities Framework.pdf. Accessed 2 Jan 2020
9. Weber LE, Duderstadt JJ (2004) Reinventing the research university. Econnomica Ltd Glion Colloquium, London
10. Weber LE, Duderstadt JJ (2006) Universities and business: partnering for the knowledge society. Econnomica Ltd Glion Colloquium, London
11. Weber LE, Duderstadt JJ (2010) University research for innovation. Econnomica Ltd Glion Colloquium, London
12. Weber LE, Duderstadt JJ (2016) University priorities and constraints. Econnomica Ltd Glion Colloquium, London
13. Crawley E, Malmqvist J, Östlund S, Brodeur DR, Edström K (2014) Rethinking engineering education: the CDIO approach, 2nd edn. Springer, Cham
14. Graham R (2018) Global state of the art in engineering education. Massachusetts Institute of Technology, Cambridge, MA

15. Cambrige-MIT Institute (2004) The Cambridge-MIT Institute (CMI). http://web.mit.edu/annualreports/pres04/13.01.pdf. Accessed 2 Jan 2020
16. CDIO (2004) The International Conceive Design Implement Operate (CDIO). In: Worldwide CDIO Initiative Standards. http://www.cdio.org/implementing-cdio-your-institution/standards. Accessed 2 Jan 2020
17. MIT Portugal Program (2006) The MIT Portugal Program (MPP). https://www.mitportugal.org. Accessed 2 Jan 2020
18. MIT and Masdar Institute Cooperative Program (2011) The MIT & Masdar Institute Cooperative Program (MIT & MICP). http://web.mit.edu/mit-mi-cp/. Accessed 2 Jan 2020
19. Singapore University of Technology and Design (2010) Singapore University of Technology and Design (SUTD): education collaboration with MIT. https://www.sutd.edu.sg/About-Us/Collaborations/MIT. Accessed 2 Jan 2020
20. van Vught FA (1999) Innovative universities. Tert Educ Manag 5:347–355
21. Neave G, van Vught FA (1991) Prometheus bound: the changing relationship between government and higher education in Western Europe. Pergamon
22. Rashdall H (2010) The universities of Europe in the middle ages: volume 1, Salerno, Bologna. Cambridge University Press, Paris
23. Arocena R, Goransonn B, Sutz J (2018) Developmental universities in inclusive innovation systems: alternatives for knowledge democratization in the global south. Palgrave Macmillan
24. Neave G, Bluckert K, Nybom T (2014) The European research university: an historical parenthesis? Palgrave Macmillan, New York, NY
25. Thelin JR (2004) A history of American higher education. Johns Hopkins University Press, Baltimore, MA
26. Geiger RL (2015) The history of American higher education: learning and culture from the founding to world war II. Princeton University Press, Princeton, NJ
27. Geiger RL (2004) To advance knowledge: the growth of American research universities, 1900–1940. Transaction Publishers
28. United Nations (2000) United Nations Millennium Declaration. https://www.un.org/en/development/desa/policy/mdg_gap/mdg8_targets.pdf. Accessed 3 Dec 2019
29. United Nations (2015) Transforming our world: the 2030 agenda for sustainable development. https://sustainabledevelopment.un.org/post2015/transformingourworld. Accessed 3 Dec 2019
30. United Nations (2000) Goal 9: Build resilient infrastructure, promote sustainable industrialization and foster innovation. In: Sustainable development goals. https://www.un.org/sustainabledevelopment/infrastructure-industrialization/. Accessed 20 Nov 2019
31. Soubbotina T (2004) Beyond economic growth : an introduction to sustainable development. The World Bank
32. Kosack S (2012) The education of nations: how the political organization of the poor, not democracy, led governments to invest in mass education. Oxford University Press
33. Vest CM (2010) Technological innovation in the 21st century. In: Weber L, Duderstadt JJ (eds) University research for innovation. Econnomica Ltd Glion Colloquium, London, pp 51–62
34. Jinping X (2014) Let engineering science and technology create a better future for humankind. In: The 2014 Annual Meeting of the International Council of Academies of Engineering and Technological Sciences (CAETS). Beijing, China
35. European Comission (2018) Horizon 2020 interim evaluation: maximising the impact of EU Research and Innovation. Brussels
36. Claudel M, Massaro E, Santi P, Murray F, Ratti C (2017) An exploration of collaborative scientific production at MIT through spatial organization and institutional affiliation. PLoS One 12. https://doi.org/10.1371/journal.pone.0179334
37. Hansen M (2009) Collaboration: how leaders avoid the traps, build common ground, and reap big results. Harvard Business Review Press, Cambridge, MA
38. Rao A, Scaruffi P (2013) A history of Silicon Valley: the greatest creation of wealth in the history of the planet. Omniware Publishing, Silicon Valley, CA

39. Aydogan N, Chen YP (2008) Social capital and business development in high-technology clusters: an analysis of contemporary U.S. Agglomerations. Springer
40. Saxenian A (1996) Regional advantage: culture and competition in Silicon Valley and Route 128. Harvard University Press, Cambridge, MA
41. Halber D (2002) Novartis is Opening Research Center in Tech Square. MIT Tech Talk
42. The Organisation for Economic Co-operation and Development (2003) The sources of economic growth in OECD countries. https://www.oecd-ilibrary.org/docserver/9789264199460-en.pdf?expires=1570827190&id=id&accname=ocid194296&checksum=F6E7CF54282C4F C489A8478B05530496. Accessed 23 Nov 2019
43. Abel JR, Deitz R (2009) Do colleges and universities increase their region's human capital? Staff Repo. Federal Reserve Bank of New York
44. Valero A, Van Reenen J (2019) The economic impact of universities: evidence from across the globe. Econ Educ Rev 68:53–67
45. Porter ME (2008) Clusters of innovation: Regional Foundations of U.S. Competitiveness. U.S. Council on Competitiveness, Washington, DC
46. Hausman N (2012) University innovation, local economic growth, and entrepreneurship. US Census Bureau Center for Economic Studies Paper No. CES-WP- 12-10
47. Porter ME (2007) Understanding competitiveness and its causes. In: Competitiveness index: where America stands. U.S. Council on Competitiveness,Washington, DC, pp 8–9
48. Moretti E (2012) The new geography of jobs. Houghton Mifflin Harcourt, Boston, MA
49. Siegfried J, Sanderson A, McHenry P (2007) The economic impact of colleges and universities. Econ Educ Rev 26:546–558
50. Samuels A (2017) Could Small-Town Harvards Revive Rural Economies? The Atlantic, Washington, DC. https://www.theatlantic.com/business/archive/2017/05/rural-economies-colleges-development/525114/. Accessed 2 Jan 2020
51. Steinacker A (2005) The economic effect of urban colleges on their surrounding communities. Urban Stud 42:1161–1175
52. Roberts E, Eesley C (2011) Entrepreneurial impact: the role of MIT—an updated report. Found Trends Entrep 7:1–149
53. Roberts EB, Murray F, Kim JD (2015) Entrepreneurship and innovation at MIT: continuing global growth and impact. Massachusetts Institute of Technology
54. Eesley CE, Miller WF (2012) Stanford University's economic impact via innovation and entrepreneurship. Stanford University
55. Pazzanese C (2015) Harvard's alumni impact. The Harvard Gazette December 8, 2015
56. United Nations (2019) Promote sustainable development. In: United Nations Promote Sustainable Development. https://www.un.org/en/sections/what-we-do/promote-sustainable-development/. Accessed 3 Nov 2019
57. Brundtland-Commission (1987) Our common future. World commission on environment and development. Oxford University Press
58. Fagerberg J (2018) Innovation, economic development and policy. Edward Elgar Publishing, Cheltenham, UK
59. Barro R, Sala-i-Martin X (2003) Economic growth. MIT Press, Cambridge, MA
60. Fitzgerald E, Wankerl A, Schramm C (2011) Inside real innovation. World Scientific Publishing Co, Singapore
61. Yusuf S, Nabeshima K (2007) How universities promote economic growth. The World Bank, Washington DC
62. Higher Education Funding Council for England (2016) University knowledge exchange (KE) framework: good practice in technology transfer. Report to the UK Higher Education Sector and HEFCE by the McMillan Group. Bristol, UK
63. Mitra J, Edmondson J (2015) Entrepreneurship and knowledge exchange. Routledge, New York, NY

64. Hughes A, Kitson M (2012) Pathways to impact and the strategic role of universities: new evidence on the breadth and depth of university knowledge exchange in the UK and the factors constraining its development. Camb J Econ 36:723–750

65. Frølund L, Murray F, Riedel M (2018) Developing successful strategic partnerships with universities. MIT Sloan Manag Rev 59:70–79

66. Byers TH, Dorf RC, Nelson A (2015) Technology ventures: from idea to Enterprise. McGraw-Hill, New York, NY

67. Mazzucato M (2013) The entrepreneurial state: debunking public vs. private sector myths. Anthem Press, London

68. Graham R (2014) Creating university-based entrepreneurial ecosystems: evidence from emerging world leaders. MIT Skoltech Initiative, Moscow, Russia. https://www.rhgraham.org/resources/MIT:Skoltech-entrepreneurial-ecosystems-report-2014-.pdf. Accessed 2 Jan 2020

69. Chan YE, Farrington C (2018) Community-based research: engaging universities in technology-related knowledge exchanges. Inf Organ 28:129–139

70. Link A, Siegel D, Wright M (2015) The Chicago handbook of university technology transfer and academic entrepreneurship. University of Chicago Press, Chicago, IL

71. Bercovitz J, Feldmann M (2006) Entrepreneurial universities and technology transfer: a conceptual framework for understanding knowledge-based economic development. J Technol Transf 31:175–188

72. Lester R (2005) Universities, innovation, and the competitiveness of local economies: local innovation systems project. MIT, Cambridge, MA

73. Lopez-Claros A (2011) The innovation for development report 2010–2011: innovation as a driver of productivity and economic growth. Palgrave Macmillan

Chapter 2
A Systematic Approach to Knowledge Exchange

2.1 Knowledge Exchange Enables Economic Development

In the first chapter, we argue that society has increasing expectations for universities' engagement in societal development. Universities already contribute broadly to cultural development, social well-being, and economic development, but more is expected. To move beyond generalizations and focus on the topic of current high expectations, we concentrate on universities' contributions to sustainable economic development.

Universities' pathway of impact on economic development is portrayed in Fig. 1.2. Economic development is influenced by the intensity of successful innovation and entrepreneurship [1]. Universities can help accelerate innovation and entrepreneurship through more effective knowledge exchange [2]. This occurs when knowledge crosses the porous boundary between universities [3] and their partners [4] in industry, small and medium enterprise [5], and government organizations [6].

This reasoning leads to our principal conclusion: **that more effective knowledge exchange with partners will likely increase universities' long-term contributions to economic development.**

This chapter develops a systematic approach to knowledge exchange in support of economic development. Section 2.2 presents a discussion of the features of bidirectional dialog with partners. Section 2.3 identifies a systematic approach that, if adapted to a particular university, would likely increase its capacity for knowledge exchange. In Section 2.4, 11 academic practices are identified that produce the outcomes used in knowledge exchange. The approach to knowledge exchange is then mapped onto the three main domains of the university—education, research, and catalyzing innovation in Sections 2.5–2.7. The chapter concludes with an assessment of how much change is needed to bring about this increased emphasis on knowledge exchange. We are mindful that the concept of knowledge exchange is also relevant for universities wanting to improve their impact on social and cultural development.

© Springer Nature Switzerland AG 2020
E. Crawley et al., *Universities as Engines of Economic Development*,
https://doi.org/10.1007/978-3-030-47549-9_2

2.2   The Essential Features of Knowledge Exchange for Economic Development

2.2.1   The University and its Partners Benefit when Knowledge Is Exchanged

Knowledge exchange occurs between a university and its partners when ideas, people, and artifacts go back and forth [7] across the porous boundaries between them [8]. Partners are stakeholders with whom universities actively exchange knowledge [9]. For a university that aims to contribute more to economic development, the main partners are industry, small and medium enterprise, and government organizations. Both the university and its partners will benefit when they cross organizational, cultural, and intellectual boundaries, sharing knowledge about their needs, capabilities, and outcomes. It is desirable that knowledge exchange occurs with some distributional equity, so that all members of society benefit.

Knowledge exchange is a multifaceted concept [10, 11] whose essential features from the perspective of universities are presented in this section and summarized in Box 2.1. Knowledge exchange will be more effective if it respects these essential features.

Box 2.1 The Essential Features of Knowledge Exchange to and from Universities in Support of Economic Development

The essential features of knowledge exchange that accelerate innovation and entrepreneurship and eventually support economic development are:
- It is influenced by the partner needs and the university's outcomes.
- It benefits from crossing disciplinary boundaries and from integrated activities in education, research, and catalyzing innovation.
- It is essentially a human endeavor, built on open communications and long-term interactions leading to trust and advocacy.
- It respects the goals, values, cultures, and priorities of the university and its partners.
- It is best guided by a systematic approach.

2.2.2   Exchanging Knowledge of Needs and Outcomes by Crossing Boundaries

Universities and their partners bring to the table important knowledge that reflects their respective needs and outcomes. The partners bring an understanding of their needs and challenges, and, more broadly, those of society. Universities bring the outcomes of their activities in the domains of education, research, and catalyzing innovation. These domains align with universities' classic missions discussed in Chapter 1: preserving and distributing knowledge, developing new knowledge, and applying knowledge. The domains correspond to the dimensions of *scholarship* identified by Boyer [12]: the scholarship of teaching, the scholarship of discovery and of integration, and the scholarship of application.

The three domains give rise to the three main outcomes of universities that support knowledge exchange with partners:

- Education yields knowledgeable and skilled university graduates and educated citizens, whom we call *talented graduates*.
- Research produces *discoveries*—new knowledge, facts, data, and theories—in single disciplines as well as across disciplines.
- Catalyzing innovations generates *creations*—the outcomes of the university that are created or synthesized: artifacts, concepts, prototypes, inventions, methods, and know-how.

Knowledge exchange is about crossing domain and disciplinary boundaries. The three domains—education, research, and catalyzing innovation—are facets of an *integrated* whole. The connections among them are as important as their internal dynamics. The integrated domains are better characterized as overlapping activities as shown in Fig. 2.1. When we identify the effective academic practices for knowledge exchange in Section 2.4, we find that many important practices transcend the boundaries among the three domains.

Universities are largely organized around traditional disciplines, which have proved a powerful influence [13], especially for systematically recording knowledge and for teaching. But these disciplinary boundaries are now under stress. The intersection of disciplines is where rich new discoveries can be found [14]. Here, new disciplines emerge that have potential importance to partners. The significant challenges of society and of industry, and of enterprise and government, do not map easily onto individual disciplines [15]. Additionally, these challenges require drawing on outcomes from multiple disciplines working together [16]. A vibrant culture of dialog *across disciplinary boundaries* facilitates exchanging knowledge with partners.

The importance of integration across domains and disciplines for effective knowledge exchange is summarized in Box 2.2. The accompanying Case 2.1 "EPFL—The Ecosystem of Education, Research, and Innovation" shows how the integration of education, research, and catalyzing innovation can foster a community of learning, curiosity, and entrepreneurship whose members work together to address societal challenges.

Box 2.2 The Principle of Integration in Knowledge Exchange

Knowledge exchange will become more effective when it is based on cross-disciplinary and integrated activities at a university—education *and* research *and* catalyzing innovation—all engaging in bidirectional dialog with partners in industry, small and medium enterprise, and government organizations.

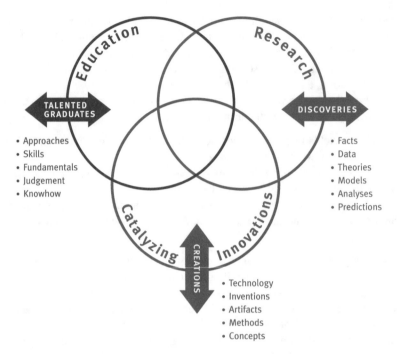

- Approaches
- Skills
- Fundamentals
- Judgement
- Knowhow

- Facts
- Data
- Theories
- Models
- Analyses
- Predictions

- Technology
- Inventions
- Artifacts
- Methods
- Concepts

Fig. 2.1 The outcomes of the three overlapping academic domains—and what knowledge is exchanged

---

Case 2.1 École polytechnique fédérale de Lausanne (EPFL)—The Ecosystem of Education, Research, and Innovation

EPFL fosters learning, curiosity, and innovation, working together to address societal challenges.

Through its diverse community of over 15,000 people, EPFL has created a spirit of curiosity, and an atmosphere of open dialog. Daily exchanges between students, researchers, and entrepreneurs encourage the emergence of new insights. At EPFL, the exchange of knowledge is linked to its three core missions: education, research, and innovation.

EPFL educates the next generation of engineers, scientists, and architects with cutting-edge degree programs. It strives to give students solid technical skills, combined with a basis in computational thinking. EPFL also encourages students to develop their imagination, creativity, and entrepreneurial spirit through cross-disciplinary projects.

EPFL offers students and faculty an innovation-oriented environment equipped with state-of-the-art facilities (see Fig. 2.2). Learning-by-doing is emphasized, and our degree programs incorporate hands-on workshops with direct research applications.

Fig. 2.2 The Rolex Learning Center, a functional and cultural hub for the EPFL ecosystem

These characteristics influence all educational programs, including the 25 master's programs currently taught at EPFL. For example, the master's program in Data Science offers a comprehensive education, from foundations to implementation, from algorithms to database architecture, and from information theory to machine learning.

The EPFL research community that is active over a spectrum of quantitative and design-focused disciplines, such as data science, personalized health, biomedical engineering, energy, robotics, and advanced manufacturing. Researchers engage in crucial societal challenges while ensuring that Switzerland remains at the cutting edge of modern technology.

As part of its research programs, EPFL has launched major strategic initiatives such as the Blue Brain Project, the Swiss Plasma Centre, and the Venice Time Machine.

The Venice Time Machine project is building a multidimensional model of Venice and its evolution throughout an entire millennium, based on the digital reconstruction of the city's administrative documents. The diversity, quantity, and accuracy of the Venetian documents are unique in Western history. The Venice Time Machine puts in operation a technical pipeline to transform this heritage of Big Data of the Past. It combines data science and document reconstruction to form an open archive for digital humanities.

EPFL has long been a key player in innovation by promoting interaction with industry, encouraging entrepreneurship and launching initiatives to strengthen collaboration with businesses through technology transfer. Innovation is pivotal to the health of the Swiss economy, where firms compete based on incorporating technology and creating new value.

The Innovation Team acts as the interface between academia and industry. Its mission is to ensure that the impact of EPFL's research is real. It connects EPFL's labs to industrial partners in a variety of ways. EPFL catalyzes discovery and technology transfer by coordinating a diverse research infrastructure so that students, researchers, and entrepreneurs are active in the exchange of knowledge and ideas.

As an example, EPFL and ETH Zurich have created a National Center for Data Science to foster innovation in open data science, aimed at making Switzerland globally competitive [1]. The center will create an open one-stop-shop for hosting, exploring, and analyzing curated, calibrated, and anonymized data.

Such cooperative strategic ventures are enabled by the governing structure. EPFL is in direct dialog with the Swiss Federal Council to ensure that EPFL's orientation is relevant for the country's objectives. EPFL is directed by the ETH Board which is appointed by the Federal Council. The Board is responsible for the implementation of the Federal Council's strategic objectives, the strategy of EPFL, and the allocation of federal funds to the institution.

Prepared based on input provided by EPFL's communication service, Mediacom, Lausanne, Switzerland.

Reference

1. EPFL (2018) Annual Report. https://www.epfl.ch/about/overview/annual-report/. Accessed 20 Jan 2020

2.2.3 The Human and Cultural Features of Knowledge Exchange

Knowledge exchange is essentially a human endeavor and requires communication and trust. In education, the main mechanism of knowledge exchange comes when new graduates enter working life—literally, humans transporting knowledge. In research and catalyzing innovation, knowledge exchange is effective when there is effective communication at all levels between counterparts at the university and partners [17]. The pathways for knowledge flow are built from a shared vision, informal interactions, and frequent open two-way discussions between counterparts [18]. These informal interactions contrast with formal mechanisms like publications, licenses, and lectures. The formal mechanisms are important, but are not, by themselves, enough for effective knowledge exchange [19].

Knowledge exchange works best when facilitated by mutually respected counterparts. In time, they gain confidence and trust in each other and their respective teams. This human trust builds confidence in the validity of the knowledge exchanged. Eventually, the counterparts become advocates for the exchange, and particularly for the adoption of the new knowledge at the partner. We return to this subject of knowledge exchange as a human endeavor in Chapter 5.

Knowledge exchange should respect and adapt to universities' and their partners' goals and cultures. The fact is that universities and their partners have differing goals, values, and priorities [20]. They operate on different timescales, funding models, and incentives [21]. In working with international partners, there may be differing national cultural norms as well [22]. To be effective, knowledge exchange has to recognize these differences and seek a mutual respect that acknowledges them [23]. University thought and administrative leaders, and their counterparts at the partners, would do well to acknowledge these differences and quickly move on to substantive relationships that will accelerate innovation and entrepreneurship.

In the discussion above, we identify some of the essential features of knowledge exchange. While parts of it can happen spontaneously, it will not become embedded in universities' thinking and actions unless it is systematically planned, operated, and incentivized. We now present a systematic approach that respects these essential features of knowledge exchange. It is set in the context of a research university that embraces innovation. The approach is a reference and not prescriptive and requires review and adaptation by each university.

2.3   A Systematic Approach to Strengthening Knowledge Exchange

2.3.1   The Three Key Actions of Systematic Knowledge Exchange

Viewing knowledge exchange in a systematic way will help us to understand and improve it. This will benefit society, the partners, and the university [24]. Knowledge exchange requires activating a diverse set of organizational interactions and relationships at the boundaries [25]. The principle of Box 2.3 provides guidance for further development of a systematic approach.

> **Box 2.3 The Principle of Systematic Knowledge Exchange**
>
> Knowledge exchange will become more effective when there is a systematic approach. Such an approach carefully identifies needs of the partners and society, conducts university activities with a sensitivity to these needs, and proactively exchanges universities' outcomes with partners.

The *systematic* approach we propose is built on the essential features defined in Box 2.1. It involves three key actions that should be adapted to each of the academic domains: education, research, and knowledge exchange. First, the university should engage partners that are particularly relevant, and that can contribute to, and gain from, engagement. The needs of society, the partners, and the university should be identified. Next, the activities of the university should be conducted with a sensitivity and responsiveness to societal and partner needs. Finally, there should be a proactive process of actually exchanging the knowledge with the partner and supporting its adoption. This process includes the migration of talented students to gainful employment [26]. These key actions are summarized in Box 2.4.

Box 2.4 The Systematic Approach to Knowledge Exchange

The proposed systematic approach to knowledge exchange contains three key actions:
- Careful identification of relevant partners in knowledge exchange, followed by a bidirectional discussion of the needs of society, the partners, and the university.
- Sensitivity and responsiveness to these needs in the conduct of education, research, and catalyzing innovation.
- A proactive process that exchanges the talented graduates, discoveries, and creation outcomes of the university and assists in their uptake by the partners.

The foundation of systematic knowledge exchange is recognizing universities' academic domains and adapting the three key actions for each. Each domain—education, research, and catalyzing innovation—has its own partners and their associated needs. Each domain has its own goals, patterns of behavior, outcomes, and exchange mechanisms (Fig. 2.1). While we describe a generic version of the systematic approach below, in reality it must be adapted to each of the domains, as discussed in Sections 2.5–2.7.

2.3.2   Identifying Partners and their Needs for Knowledge

Partners are those stakeholders with whom universities actively exchange knowledge. These partners can be in industry, small and medium enterprise, government organizations, as well as in nonprofits, civic organizations, and other types of institutions. Partners should have shared interest with the university. They should be willing to disclose information regarding their needs. It is best if they are also in a position to make constructive use of a university's outcomes.

To be able to guide its activities towards relevant outcomes, a combination of university scholars, academic units, and university leadership should try to understand the partners' plans, as well as their problems, opportunities, and challenges. This can be done by individual scholars in an informal way. Or it can be done by engaging in a more formal bidirectional dialog in which the university and its partners talk and listen carefully. Sometimes facilitation can support such dialog, and often several cycles are needed to reach convergence and understanding. This dialog step is relatively similar in the domains of education, research, and catalyzing innovation. We return to this point in the practice of Facilitating Dialog and Agreements in Chapter 5.

An example of this systematic approach to identifying partners and their needs is presented in Case 2.2 "Skoltech—Systematic Knowledge Exchange to Establish Research." Here, a start-up university deeply engages both industrial partners and scientific peers to outline a strategy for research investment. The strategy is designed to satisfy the curiosity of scholars and the desire for impact by government stakeholders.

Case 2.2   The Skolkovo Institute of Science and Technology (Skoltech)—Systematic Knowledge Exchange to Establish Research

A systematic approach to knowledge exchange was the foundation of Skoltech and accelerated its development as a research and innovation university.

At its founding in 2011, the Skolkovo Institute of Science and Technology (Skoltech) was envisioned as a graduate university focused on information technology, energy, biomedicine, and other high technology sectors pertinent to the Russian economy. Its mission was to accelerate innovation by developing relevant cutting-edge science and technology, and moving it quickly and effectively from the academic to the commercial communities.

The objective of the founding team was to identify and successfully implement research, education, and innovation opportunities that yielded skilled graduates, research, and innovation outcomes, supported the growth of Skoltech, and had a significant impact on Russia.

The founders took a systematic approach to knowledge exchange to setting the initial research priorities for the university. The approach included three factors that are described in more detail below (see Fig. 2.3). The outcome set the stage for a second systematic approach for establishing the corresponding set of Centers for Research, Education, and Innovations (CREI).

The needs of the market and industry were derived from the desire to create new or improved products or services in the market, and from industry's desire to improve its organizational structure and advance manufacturing. These needs were captured through interviews with technical and marketing leaders in large multinational companies active within Skoltech's focus areas, and similar interviews with leaders of the Skolkovo Foundation's Innovation Clusters, which reflected small- and medium-sized Russian enterprise. In addition, MIT's Industrial Liaison Program surveyed several of its member companies. In total, about 35 executives and cluster leaders participated. Once the topical areas were identified, an extended set of focused workshops with Russian and international industry were held.

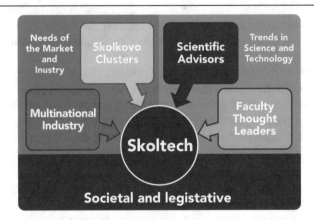

Fig. 2.3 The key influences on the research foresight study

Forecasting trends in scientific and technical development was done by engaging about 70 thought leaders and scientific advisors in Russian universities and at MIT over a period of 3 months. These individuals were sent a questionnaire tailored to Skoltech focus areas.

Society and legislative megatrends were identified and analyzed to provide insights into what needs for new solutions and new technology can be expected from, for example, urbanization, aging population, and climate regulation. This work was done as a 2-month project by a group of MBA students at the MIT Sloan School of Management.

At the end, all results were presented in a workshop format. Then, the Skoltech leadership debated the results, taking into consideration the expected deliverables, and the ability to involve a broad set of Russian institutions and companies. They then arrived at an initial Research Investment Strategy which represented a balanced portfolio close to long-term impact of its outcomes. At a high level, the research areas are: biomedicine (including infectious diseases and regenerative medicine); energy (hydrocarbon production, electrical power systems); information technology (machine learning, quantum physics); space; and cross-cutting issues (advanced materials, computationally and data intensive science).

The strategy was published widely for maximum transparency. It was then used to start hiring faculty of world-class stature interested in working in an interdisciplinary mode in the designated research directions.

As a result of applying systematic approaches, in less than 5 years Skoltech developed into a thriving research university with nine CREIs. These CREIs are starting to have a large impact on both academia and in the economy. For example, the number of papers per Skoltech faculty member in the Nature Index Journals is now comparable to KAIST.

Contributed by Dr. Mats Nordlund, Founding Vice President of Research Programs, Skoltech, Moscow, Russia.

2.3.3   Conducting the University Activity with Sensitivity and Responsiveness

To better provide for successful knowledge exchange, academic practices should be conducted with a sensitivity and responsiveness to the needs of the partners and society. University scholars and leaders should balance partners' suggestions with their own internal compass. This is not an issue of the partner interfering in university internal affairs: it is about universities learning from all possible sources before making decisions. For example, the curriculum should reflect the need for fundamental knowledge, but also the needs of society, industry, and enterprise for skills. Thus, the curriculum evolves over time. The outcomes of catalyzing innovation are tightly coupled to the needs of industry and enterprise. Even research at the frontier of knowledge can be inspired by the partners' needs.

2.3.4   Proactively Exchanging Knowledge and Advocating its Uptake

Knowledge that is successfully exchanged is the value that universities provide to society. In general, the exchange will be more successful if universities are *proactive* in their execution. Likewise, it will be more impactful if the partners *advocate* for its uptake within their organizations.

The specific content of the knowledge exchange depends on the academic domain (Fig. 2.1). Each domain has a proactive knowledge exchange process. Education prepares talented graduates with an array of knowledge and skills. Research discovers new facts, data, and theories. The creations of catalyzing innovation include technology, inventions, artifacts, methods, and concepts.

The proactive mechanism at the university also depends on the domain (Fig. 2.4). Education transfers knowledge when students leave the university, through internships, employment or when they start ventures. Research disseminates discoveries through publications and personal interactions. Similar mechanisms are used in catalyzing innovation, but additional mechanisms include intellectual property, start-ups, and consulting.

Partners are proactive when they advocate for the adoption of the university outcomes. Effective partners employ scouts who stay aware of the outcomes of universities. Such scouts include human resources specialists for graduates, internal researchers for discoveries, and product and system developers for innovation creations.

There are numerous specific models for knowledge exchange, depending on academic domain and context. We identify the broad generalities in this chapter and offer greater detail in Chapters 3–6.

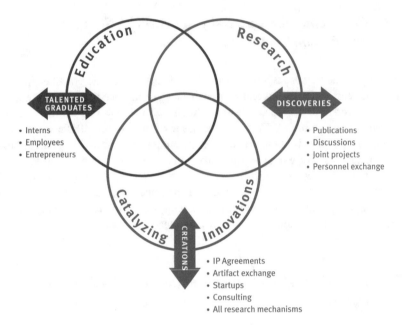

Fig. 2.4 The outcomes of the three overlapping academic domains—and how knowledge is exchanged

2.4   The Effective Academic Practices that Yield Knowledge Outcomes

2.4.1   Arriving at the Academic Practices Involved in Knowledge Exchange

We identify a set of academic practices that collectively yield the outcomes that are exchanged by universities with partners. These effective practices are based on observed patterns of behavior at universities. The practices are *effective* when they produce outcomes that cross the universities' boundaries and lead to meaningful contributions to society. From the perspective of economic development, a practice is effective when it exchanges knowledge that accelerates innovation and entrepreneurship.

We arrived at this set of practices by reflecting on behavior at multiple universities and in many projects. We asked the questions:

- What are the practices needed to represent the university fully, in its three academic domains and in their overlaps?
- What are the practices that account for economically useful outcomes?
- What can we learn from previous attempts to build frameworks that describe education, research, and catalyzing innovation?

Our framing of these practices is not unique—others could observe the same behaviors and identify a slightly different set of practices. Like all models, its value lies in its utility.

The important criterion for selecting the set of academic practices is that each should support the systematic approach, as discussed above. There should be identification of partners and their needs, as well as a sensitivity to these needs in the conduct of the practice. There should be a proactive process of knowledge exchange and uptake.

2.4.2 The Eleven Effective Academic Practices

We distinguish eleven effective academic practices that meet these criteria. They map to the overlapping academic domains as shown in Fig. 2.5 and explained in Table 2.1. The table shows the full and short name of each practice, as well as a brief description of the associated activities and their outcomes. We realize that the crisp boundaries shown in Fig. 2.5 are in reality more fuzzy, but showing these internal boundaries is useful for the analysis.

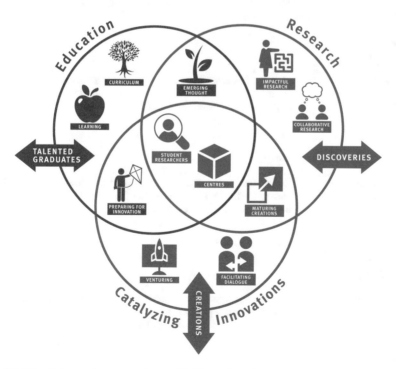

Fig. 2.5 The eleven academic practices: red indicates the primary outcomes are talented graduates, blue indicates discoveries, and green creations

Table 2.1 The effective academic practices that yield the outcomes used in knowledge exchange

Practice name and icon	Description of the practice > and its outcome
Integrated curriculum CURRICULUM	Implementing an integrated curriculum of courses, projects, and co-curricular experiences > educating students who learn the disciplinary fundamentals together with the essential skills, approaches, and judgment
Teaching for learning LEARNING	Engaging students in active, experiential, and digital learning > graduating students with deep conceptual understanding, self-efficacy, and capability for self-learning
Education in emerging thought EMERGING THOUGHT	Quickly migrating cross-disciplinary and emerging research outcomes to the curriculum and learning opportunities > preparing students who learn emerging disciplines, technologies, and bodies of thought
Preparing for innovation PREPARING FOR INNOVATION	Implementing educational activities for leadership, management, and entrepreneurship > educating students who are better prepared for their potential roles in innovation
Impactful fundamental research IMPACTFUL RESEARCH	Pursuing fundamental new discoveries along a spectrum from curiosity-driven to use-inspired > yielding new knowledge with an impact on scholarship and society
Collaborative research within and across disciplines COLLABORATIVE RESEARCH	Collaborating with other scholarly researchers within the universities and externally > making discoveries that cross-disciplinary boundaries and are in new fields of thought
Centres of research, education and innovation CENTRES	Empowering larger scale integrated Centres of Research, Education, and Innovation > producing directly implementable and impactful solutions that address significant issues of society

(continued)

Table 2.1 (continued)

Practice name and icon	Description of the practice > and its outcome
Undergraduate and postgraduate student researchers STUDENT RESEARCHERS	Engaging undergraduate and postgraduate students in research > energizing research, and yielding effective young researchers and agents of knowledge exchange
Maturing discoveries and creations MATURING CREATIONS	Making progressive discoveries, creations, inventions, market analyses, and proof-of-concept demonstrations > yielding creations with higher technology and market readiness
Facilitating dialog and agreements FACILITATING DIALOGUE	Actively facilitating informal dialog and formal agreements with partners > improving understanding of partner needs and enabling more university creations to be adopted by partners
University-based entrepreneurial venturing VENTURING	Engaging in the real entrepreneurial process within the university, supported by networks of mentors, with access to investors and facilities > producing new ventures and more experienced entrepreneurs

As suggested by Fig. 2.5, there are practices such as Teaching for Learning that are largely in one domain. There are practices such as Education in Emerging Thought that bridge two domains. And in the three-way overlap, we find Centres of Research, Education, and Innovation. These Centres apply scientific research outcomes to problems of innovation, all while engaging students.

In this chapter, we present an overview of the 11 academic practices in this systematic approach. The practices themselves are discussed in Chapters 3–5.

2.5   Education as a Systematic Approach to Knowledge Exchange

2.5.1   Education, its Partners and their Needs

Education is about learning [27]. Under the guidance of instructors, students acquire deep working knowledge of disciplinary fundamentals and essential skills, approaches, and judgment, better preparing them to act as agents of societal development. Instructors also learn and develop new knowledge as they prepare and reflect. Upon graduation, students become important agents of knowledge exchange and carry their learning with them into their next phase of life.

Viewed through the lens of economic development, the specific objective of education is to prepare students with a *deep working knowledge of fundamentals while preparing them to be more effective agents of knowledge exchange and innovation.* They might eventually contribute through careers specifically in innovation or entrepreneurship, or more broadly in education, research, management, public policy, or environmental protection.

The relevant partners in education are the future employers of graduates in industry, small and medium enterprise, and government organizations. The needs they report speak to disciplinary knowledge—for example, they need more IT experts—and also to essential skills—such as teamwork. The government is, broadly, a stakeholder in education, as are professional bodies. Partners' needs are often reflected in national standards, or in desired learning outcomes set by the quality agencies or professional bodies.

How scholars learn about partners' needs is guided by the educational practice of the Integrated Curriculum. This involves the faculty setting desired learning outcomes that are informed by stakeholder input. This helps ensure that students acquire the appropriate foundations that will enable them to create value for themselves, their employers, and society. The practice calls for engaging with partners as well as other key stakeholders (e.g., graduates, faculty, students, professional bodies) to help define desired learning outcomes (Chapter 3).

2.5.2   A Responsive Approach by Universities to Needs in Education

The goals of education can be addressed by adapting the four practices whose outcomes are talented graduates. These practices are listed in Table 2.1 and in Fig. 2.5, which suggests whether the practice is largely in the domain of education, or in overlaps with research or innovation. The goals will be met when engaged and responsive faculty members:

- Equip students with a strong foundation in fundamentals and skills, approaches, and judgment, employing the practice of Integrated Curriculum.
- Develop in students deep working understanding, self-efficacy, and the capability for self-learning, using Teaching for Learning.

- Extend students' learning to include emerging disciplines, technologies, and bodies of thought, using Education in Emerging Thought.
- Prepare students to take roles as young professionals in innovation by engaging in Preparing for Innovation.

These practices are summarized in Table 2.1 and are discussed in greater depth in Chapter 3. We have to keep in mind that students' learning is not only shaped by the curriculum, but also by the broader student experience, and by employment during their university years.

2.5.3   A Proactive Process to Exchange Educational Outcomes

Students and graduates are important mechanisms of knowledge exchange. When they enter the world of work, they carry with them knowledge: a deep understanding of the fundamentals in their field of study. They also bring essential skills, approaches, and judgment (Fig. 2.1). They may also have acquired research and innovation know-how, and they may have learned about emerging fields of thought. These capabilities make students potentially important contributors to accelerating innovation.

This knowledge is exchanged between the university and partners when students participate in internships and associate with industrial mentors (Fig. 2.4). The most direct and important knowledge exchange occurs when the talented graduates leave the university to start a new venture, or take up employment as an innovator or policy maker.

The proactive process occurs when the university works to place the students where they can contribute. Partner advocacy occurs when the new employer arranges a relevant and meaningful position where the graduates can productively make use of their knowledge, skills, and know-how.

2.6   Research as a Systematic Approach to Knowledge Exchange

2.6.1   Research, its Partners and their Needs

Research is the discovery of new knowledge at the frontiers, and the quest for increased understanding of our world. The general objective of research is to make *discoveries*, often revealing phenomena or truths that have previously existed but were unknown or unexplained. We use the word *discoveries* to distinguish research outcomes from the synthesized outcomes of catalyzing innovation that we call creations (see below). We note that new knowledge also can be developed in the course of education and catalyzing innovation.

Viewed through the lens of economic development, the specific objective of research is *to make discoveries at the frontiers of knowledge, that have the potential for becoming more effective instruments of knowledge exchange and innovation.*

The motivation for research ranges from curiosity-driven to use-inspired research, and on to research that creates a directly implementable solution. Faculty have the right and responsibility to select problems important to them and produce impactful results (Chapter 4). One feature that distinguishes curiosity-driven research from use- or impact-inspired work is the degree of external engagement at the time of formulating the research question.

In curiosity-driven research, scholars are motivated by interesting problems at the frontiers of knowledge, which may or may not be immediately relevant to existing societal or industry issues. They can benefit from collaboration with like-minded scholars, but otherwise external engagement may be low. This kind of research advances knowledge and can often lead to unexpected discoveries with far-reaching applications.

Use-inspired research also seeks discoveries at the frontier, but is motivated by problems of industry or society. It can also lead to unexpected fundamental discoveries and consequent peer recognition. In use-inspired research, scholars scan their world for addressable issues of a fundamental nature. When an issue is found, the scholars might engage with appropriate external counterparts to develop a mutual understanding.

A third type of research is chartered from the beginning to develop directly implementable solutions to larger scale problems of industry, enterprise, government, and society. Usually conducted by a cohesive team that includes scholars and partners, the formulation of the research question is intrinsic to the process and often influenced by those with the needs.

Research groups and universities would do well to have a balanced portfolio of these approaches. This balance will create knowledge outcomes that will influence economic development in the near-, mid-, and long term.

2.6.2   A Responsive Approach by the University to Needs in Research

There are four practices that support the goals of research with potential impact on innovation (Table 2.1). The placement of the practices in Fig. 2.5 suggests that the practice is either in the domain of research or sits in the total overlap area. The goals are more likely to be met when the dedicated researchers commit to:

- Increasing the scholarly and innovation *impact* of research discoveries using the practice of Impactful Fundamental Research.
- Broadening the *scope* of research, working across disciplinary boundaries and in new fields of thought by engaging in the practice of Collaborative Research.

- Developing *directly implementable and impactful solutions* that address significant issues of society by deploying the practice of Centres of Research, Education, and Innovation.
- Invigorating discoveries, and training future researchers and agents of knowledge exchange by using the practice of Undergraduate and Postgraduate Student Researchers.

Table 2.1 summarizes these practices, and they are considered in significantly more detail in Chapter 4.

2.6.3   A Proactive Process to Exchange Research Outcomes

Research yields new knowledge, explains mysteries, and inspires young and old. Research influences the curriculum, helps educate students, and contributes to innovative creations. Proactive knowledge exchange of research outcomes also spans this spectrum. The outcomes of research are the discovery of knowledge, facts, data, and theories that better explain our world. Research also yields new models, analyses, and predictions (Fig. 2.1).

Knowledge exchange around discoveries usually combines formal publications with informal discussions (Fig. 2.4). These may lead to sharing of detailed results with scholar-partners and to joint projects. Involvement of researchers in presentations, professional education, and personnel exchange is also important.

Proactive knowledge exchange requires the scholar's continued engagement in dissemination after the discovery is made and first presented. External advocacy occurs when peers recognize and cite the contribution, and when partners attempt to apply it.

Research is increasingly a cross-disciplinary endeavor [28]. To produce the outcomes for external knowledge exchange, scholars within the university must be prepared to cross-disciplinary boundaries as well. This is discussed in Box 2.5.

Box 2.5   Crossing Boundaries among Disciplines

Universities with an aspiration to strengthen economic development have to consider how they can best organize themselves to meet both traditional missions and the new expectations of society. We propose an activity-based approach to addressing cross-disciplinary challenges.

From the list of 11 effective academic practices, two are explicitly about crossing boundaries between disciplines. Education in Emerging Thought brings new cross-disciplinary research outcomes into the curriculum. Collaboration Within and Across Disciplines suggests new approaches to cross-disciplinary research.

Four of the other practices directly take on the challenges of partners and society. Such challenges rarely respect disciplinary boundaries and are therefore implicitly cross-disciplinary:
- Maturing Discoveries and Creations.
- Facilitating Dialog and Agreements.
- Centres of Research, Education, and Innovation.
- University-Based Entrepreneurial Venturing.

When students and scholars routinely cross internal disciplinary boundaries, universities will potentially contribute more to knowledge exchange:
- Discoveries that integrate between traditional disciplines.
- Breakthrough opportunities at the boundaries of disciplines.
- Access to unique data and varied perspectives on its interpretation.
- Students prepared to work across disciplinary boundaries.

2.7   Catalyzing Innovation as a Systematic Approach to Knowledge Exchange

2.7.1   Catalyzing Innovation, its Partners and their Needs

Catalyzing innovation is universities' newest role, and some believe it is as important as education and research [29–31]. The general objective of catalyzing innovation is to produce *creations*: objects, processes, and systems that have never before existed, and that have potential for societal impact. These can range from inventions and technologies to business models, healthcare solutions, and works of art.

Through the lens of economic development, the goal is to *more effectively stimulate and capture the richness of innovation creations in the university, and exchange them with industry, enterprise, and government organizations.* Once exchanged, creations flow to products, services, and systems that are brought to market or otherwise deployed by partners.

The partners in catalyzing innovation are those who stand to benefit from the creations. The boundary between these partners and the university benefits from being fuzzy. In a well-functioning innovation ecosystem, there are many participants from both sides who hug closely to the boundary, and cross it from time to time. Knowledge exchange benefits from this fuzziness and porosity.

Means of gaining an understanding of partner needs in catalyzing innovation range from informal discussion, often enabled by networking, to structured, and professionally led dialog. One structured process that can be applied in all domains is called "systematic dialog." Here, thought leaders from the university and partners engage in discussion with sufficient time and facilitation to allow improved sharing of ideas (Chapter 5).

2.7.2   A Responsive Approach by Universities to Needs in Catalyzing Innovation

The remaining three practices of Table 2.1 support the goals of catalyzing innovation for economic development. Figure 2.5 places these practices in the zones of catalyzing innovation and overlapping with research. An engaged collective of faculty, staff, and students will contribute to meeting these goals by:

- Raising the technology and market readiness of university innovation creations using the practice of Maturing Discoveries and Creations.
- Engaging with partners, understanding their needs and championing their effective adoption of university creations by applying the practice of Facilitating Dialog and Agreements.
- Supporting real venture creation by faculty, staff, and students using the practice called University-Based Entrepreneurial Venturing.

These practices are summarized in Table 2.1 and are explored more thoroughly in Chapter 5.

2.7.3   A Proactive Process to Exchange Innovation Outcomes

Universities can play an important role in catalyzing innovation by better preparing students to be innovators and entrepreneurs [32], by better maturing creations to reduce barriers to adoption [33], and by providing more support to spinoffs [34]. All contribute to knowledge exchange.

The creations that are exchanged with partners include technologies, inventions, artifacts, methods, and concepts [35]. Other creations exchanged are intellectual property (e.g., copyrights, trademarks), tangible research property (e.g., engineering prototypes, drawings, new organisms, software, circuit chips), know-how, and business ideas [36] (Fig. 2.1).

These can be exchanged by the same mechanisms associated with research: publications, discussions, joint projects, and personnel exchange. Distinctive mechanisms span intellectual property and tangible research property agreements, exchange of tangible artifacts, and involvement in start-ups and consulting (Fig. 2.4). In addition, there are events, shows, and various networks.

Proactive knowledge exchange happens when a university creator works closely with the partner to support uptake and implementation. Advocacy is evident when those in the partner organization push to build the creation into a product or system.

Catalyzing innovation is an activity that inherently overlaps with education and research. In fact, universities produce a great deal of value in the areas of overlap among the three domains. Box 2.6 discusses these important interactions.

Box 2.6 Crossing Boundaries among Education, Research, and Catalyzing Innovation

Universities are systems with three academic domains: education, research, and catalyzing innovation. Their interaction enables important outcomes to emerge. Effective knowledge exchange depends on the overlap of the domains as suggested in Fig. 2.5. We identify practices in each of the overlaps:

- Research interacts with education when research results move quickly into educational programs through Education in Emerging Thought.
- Research feeds ideas to innovation and innovation's needs influence research through Maturing Discoveries and Creations.
- Education and innovation overlap when education deals with leadership, management, and entrepreneurship issues, providing highly skilled graduates for innovation and entrepreneurial ventures through the practice of Preparing for Innovation.
- The triple overlap of education, research, and catalyzing innovation solves impactful problems through the practice of Centres of Research, Education, and Innovation. It engages students in authentic research and innovation through the practice of Undergraduate and Postgraduate Student Researchers.

By crossing these internal domain boundaries, universities will have more to contribute to partners and society:

- Graduates who are at the cutting edge of emerging knowledge and experienced in research and innovation.
- Research discoveries that are fundamental, and perhaps are on a path to innovation impact.
- Creations that reflect the enthusiasm and creativity of students and are built on the newest discoveries.

2.8   Summary and the Change Needed to Strengthen Knowledge Exchange

Universities will not be surprised by the need for change [37], nor should they be fearful of it [38]. Change in universities is both necessary and possible. Necessary, because of the expanding aspirations in economic development and the quickening pace of society's expectations. Possible, because universities have a long history of constructive adaptation, as discussed in Chapter 1. But the pace of change is quickening. Change is an ongoing process at universities.

We have confidence that universities' engagement with society can be strengthened. This can be done by considering the essential features of knowledge exchange (Section 2.2) and following a systematic approach, as discussed in (Section 2.3). These features and approaches support the effective practices (Sections 2.4–2.7).

The practices reinforce the traditional value of education and research, along with the newer domain of catalyzing innovation. The systematic approach is both a call to strengthen universities' fundamental mission and a call for significant change in how it is carried out.

Many universities have already engaged in aspects of knowledge exchange, driven by individual action and strategic intent. Few can claim that they have done enough since the goalposts are continually moving. Despite these efforts by universities, there are significant gains still to be made. Even for leading research universities well-known for embracing innovation, significant improvements are possible. Imagine the influence of a university on the world if:

- Every graduating student possesses the knowledge and skills to be a successful innovator or entrepreneur.
- Every discovery is quickly examined for its potential impact, and a translational plan created for its development and dissemination.
- Every applicable creation is patented, licensed, and embedded in a marketable product, service, or system.

The degree to which universities can contribute depends very much on the university, its context, its level of ambition, and amount of action already taken. For emerging universities starting with a modest level of engagement with economic or societal development, adapting our practices would represent a great but worthwhile challenge. For new universities that are being planned, there is an opportunity to build a knowledge exchange philosophy from the start. For the majority of universities that are well established and developed, the degree of change necessary can only be revealed by an honest process of self-assessment.

No one simple change will make a university a superior performer in knowledge exchange. It requires a process of continuous improvement, and an extended commitment by university leaders, scholars, and partners to work together. *The two questions governing change are*: How much is needed, and how fast must it take place?

To provide support for change, we describe effective practices for education, research, and catalyzing innovation in Chapters 3–5, and take an integrated view in Chapter 6. In Chapters 7–10, we lay out a framework for the university as an adaptable organization. This includes supporting practices, program evaluation, alignment with partners, and an expanded discussion of change. Nothing in these effective academic practices or the framework for an adaptable university is intended to be prescriptive. We offer a resource, reference, and examples of behaviors that should be examined and adapted by each university to its local needs and context [39].

References

1. Shane S (2006) Economic development through entrepreneurship: government, university and business linkages. Edward Elgar Pub
2. Mitra J, Edmondson J (2015) Entrepreneurship and knowledge exchange. Routledge, New York, NY

3. The Organisation for Economic Co-operation and Development (2012) A guiding framework for entrepreneurial universities. https://www.oecd.org/site/cfecpr/EC-OECD Entrepreneurial Universities Framework.pdf
4. Langfitt T, Hackney S, Fishman AP (1983) Partners in the research enterprise. University of Pennsylvania Press, Philladelphia, PA
5. The Organisation for Economic Co-operation and Development (2015) The innovation imperative: contributing to productivity, growth and well-being. Organization for Economic Cooperation and Development, Paris, France
6. Szirmal A, Naude W, Goedhuys M (2011) Entrepreneurship, innovation, and economic development. Oxford University Press
7. Hughes A, Kitson M (2012) Pathways to impact and the strategic role of universities: new evidence on the breadth and depth of university knowledge exchange in the UK and the factors constraining its development. Camb J Econ 36:723–750
8. Yusuf S (2008) Intermediating knowledge exchange between universities and businesses. Res Policy 37:1167–1174
9. Mitra J (2012) Entrepreneurship, innovation, and regional development. Routledge, London
10. Abreu M, Grinevich V, Hughes A, Kitson M, Ternouth P (2008) Universities, business and knowledge exchange. Council for Industry and Higher Education and the Centre for Business Research at University of Cambridge
11. Agrawal A (2001) University-to-industry knowledge transfer: literature review and unanswered questions. Int J Manag Rev 3:285–302
12. Boyer EL (1990) Scholarship reconsidered: priorities of the professoriate. The Carnegie Foundation for the Advancement of Teaching, Princeton, NJ
13. Jacobs J (2013) In Defense of disciplines: interdisciplinarity and specialization in the research university. Chicago, IL, The University of Chicago Press
14. Gethmann CF, Carrier M, Hanekamp G, Kaiser M, Kamp G, Ligner M, Quante M, Thiele F (2014) Interdisciplinary research and trans-disciplinary validity claims. Springer
15. Libecap G, Thursby M, Hoskinson S (2010) Spanning boundaries and disciplines: university technology commercialization in the idea age. Advances in the study of entrepreneurship, innovation and economic growth. Emerald Group Publishing
16. O'Brien L, Marzano M, White R (2013) Participatory Interdisciplinarity: towards the integration of disciplinary diversity with stakeholder engagement for new models of knowledge production. Sci Public Policy 40:51–61
17. Clare P (2018) Knowledge exchange. In: Andersen J, Toom K (eds) Research management. Elsevier, pp 189–203
18. Niedergassel B (2011) Knowledge sharing in research collaborations: understanding the drivers and barriers. Springer
19. Chan YE, Farrington C (2018) Community-based research: engaging universities in technology-related knowledge exchanges. Inf Organ 28:129–139
20. Ackworth EB (2008) University–industry engagement: the formation of the knowledge integration community (KIC) model at the Cambridge-MIT Institute. Res Policy 37:1241–1254
21. Belenzon S, Schankerman M (2009) University knowledge transfer: private ownership, incentives, and local development objectives. J Law Econ 52:111–144
22. The Organisation for Economic Co-operation and Development (2019) University-Industry Collaboration: New Evidence and Policy Options. https://read.oecd-ilibrary.org/science-and-technology/university-industry-collaboration_e9c1e648-en#page1. Accessed 1 Feb 2020
23. Martin LM, Warren-Smith I, Lord G (2019) Entrepreneurial architecture in UK universities: still a work in progress? Int J Entrep Behav Res 25:281–297
24. Higher Education Funding Council for England (2016) University knowledge exchange (KE) framework: good practice in technology transfer. Report to the UK Higher Education Sector and HEFCE by the McMillan Group. Bristol, UK
25. Zhang Q (2018) Theory, practice and policy: a longitudinal study of university knowledge exchange in the UK. Ind High Educ 32:80–92

26. Borrell-Damian L, Brown T, Dearing A, Font J, Hagen S, Metcalfe J, Smith J (2010) Collaborative doctoral education: university-industry partnerships for enhancing knowledge exchange. High Educ Pol 23:493–514

27. Ackoff R, Greenberg D (2008) Turning learning right side up: putting education Back on track. Pearson Prentice Hall, Upper Saddle River, NJ

28. Atkinson J, Crowe M (2006) Interdisciplinary research: diverse approaches in science, technology, health and society. John Wiley & Sons, Hoboken, NJ

29. Leydesdorff L, Etzkowitz H (1997) Universities and the global knowledge economy: a triple helix of university-industry-government relations (science, technology, and the international political economy series). Thomson Learning, Boston

30. Roberts EB (1991) Entrepreneurs in high technology: lessons from MIT and beyond. Oxford University Press, New York, NY

31. Allen T, O'Shea RP (2014) Building technology transfer within research universities: an entrepreneurial approach. Cambridge University Press, Cambridge, UK

32. Valerio A, Parton B, Rob A (2014) Entrepreneurship education and training programs around the world. World Bank Publications, Washington, DC

33. Shane S (2004) Academic entrepreneurship: university spinoffs and wealth creation. Edward Elgar Pub

34. Birley S (2007) Universities, academics, and spinout companies: lessons from Imperial. Int J Entrep Educ 9:388–408

35. Breznitz S, Etzkowitz H (2016) University technology transfer: the globalization of academic innovation. Routledge

36. Libecap GD (2005) University entrepreneurship and technology transfer: process, design, and intellectual property. Elsevier (Series Advances in the Study of Entrepreneurship, Innovation, and Economic Growth)

37. Altman A, Ebersberger B (2013) Universities in change: managing higher education institutions in the age of globalization. Springer, New York, NY

38. Benson L, Harkav I, Puckett J, Hartley M, Hodges R, Johnston F, Weeks J (2017) Knowledge for social change: bacon, Dewey, and the revolutionary transformation of research universities in the twenty-first century. Temple University Press

39. Altbach PG (2011) The past, present, and future of the research university. In: Altbach PG, Salmi J (eds) The road to academic excellence. World Bank Publications, pp 11–32

Chapter 3
Education and Knowledge Exchange

3.1 Introduction

3.1.1 The Objectives of Education

Education is the main purpose for which universities were once established, and still the mission that distinguishes them from other types of research and innovation institutes. Education is also arguably the most valuable contribution of the university to society. The role of the university in education is to create a learning environment that enables students to develop knowledge and essential skills. Students come to the university and invest their time and energy to prepare themselves for work and life. Then they leave as talented graduates, with all their capabilities, energy, and aspirations.

Box 3.1 Objectives of Education and Knowledge Exchange

The general objective of education is to develop the potential of the students to lead fulfilling, productive lives. They become *talented graduates* who are able to contribute to knowledge exchange and more broadly to society. The outcomes of their education include deep working understanding of established fundamentals and emerging knowledge, and essential life and professional skills, including know-how in research and innovation (Fig. 2.1).

Knowledge is exchanged when students work as interns, and when they leave the university for employment or to start new ventures (Fig. 2.4).

Viewed through the lens of economic development, the specific objective of education is to **prepare students with a deep working knowledge of fundamentals while better preparing them to be more effective agents of knowledge exchange**

© Springer Nature Switzerland AG 2020
E. Crawley et al., *Universities as Engines of Economic Development*,
https://doi.org/10.1007/978-3-030-47549-9_3

and innovation. Graduates might eventually have careers in industry and enterprise, as well as in education, research, public policy, and protection of the environment.

Chapter 2 argues that education makes important contributions to economic development. The line of reasoning is that *economic development* on the macro-level is supported by accelerating innovation and entrepreneurship at the enterprise level. This is assisted by *knowledge exchange* between the university and industry, small and medium enterprise and government organizations. One principle of Chapter 2 is that an *integrated system of activities* at a university has the greatest potential to substantially enhance knowledge exchange and accelerate innovation. The second principle is that knowledge exchange will be more effective when guided by a systematic process, that considers partner needs, and proactively exchanging outcomes. We advocate a systematic process that leads to constructive interplay of education *and* research *and* innovation, all engaging with partners outside the university. This chapter focuses on the educational component of this systematic, integrated set of activities. This component is not isolated from research and catalyzing innovation, but interacts with them considerably.

3.1.2   Background and Opportunity

Our message is that a university that seeks to expand its mission of contributing to sustainable economic development should seriously reexamine and invest in student learning. Without detracting from any of the present value that education creates, we see a key opportunity in realizing the full potential of education *for knowledge exchange*.

Students need to develop *deep working knowledge* of disciplinary fundamentals infused with the skills, approaches, and judgment that enable them to access both existing and new opportunities. They need to develop essential life and professional skills, including know-how in research and innovation. When graduates are well prepared for a productive working life it will benefit the graduate, their employers, and a sustainable society. Their successes will also influence the reputation of the university.

Most universities have well-established and ambitious educational goals. Faculty are deeply engaged in fulfilling this mission, and students do graduate to become successful contributors in society and industry. We find, however, that education is not yet fully understood as arguably the *most* important enabler of economic development, through the large-scale knowledge exchange by graduates.

3.1.3   The Academic Practices of Education

This chapter focuses on four effective academic practices that strengthen knowledge exchange through education. Summarized in Table 3.1, these four effective academic practices are discussed in more detail in the following sections.

Table 3.1 The educational practices

Practice icon	Practice name: description of the practice > and its outcome
CURRICULUM	*Integrated curriculum*: Implementing an integrated curriculum of courses, projects, and co-curricular experiences > educating students who learn the disciplinary fundamentals together with essential skills, approaches, and judgment.
LEARNING	*Teaching for learning*: Engaging students in active, experiential and digital learning > graduating students with deep working understanding, self-efficacy, and capability for self-learning.
EMERGING THOUGHT	*Education in emerging thought*: Migrating cross-disciplinary and emerging research outcomes quickly to the curriculum and learning opportunities > preparing student who learn emerging disciplines, technologies, and bodies of thought.
PREPARING FOR INNOVATION	*Preparing for innovation*: Implementing educational activities for leadership, management, and entrepreneurship > educating students who are better prepared for their potential roles in innovation.

Interspersed are 12 case studies that contribute to and illustrate the practices. In Chapter 6, these educational practices are integrated with those in research and catalyzing innovation.

The potential influence of these practices on the university and its faculty is indicated by the transition from a reference situation at a good university to an aspirational one, summarized in Table 3.2.

We propose these four educational practices as a set of coordinated strategies for working towards the aspirational situation. There is, however, a difference between the first two and the second two practices. *Integrated Curriculum* and *Teaching for Learning* describe the sound educational development that most universities could strive for. We argue that when a university considers how education can contribute to sustainable economic development these practices become truly crucial. The next two practices, *Education in Emerging Thought* and *Preparing for Innovation*, clearly address more specific aspects of preparing for knowledge exchange and

Table 3.2 Reference and aspirational situation for education

Reference situation for education	Aspirational situation for education
The intended learning outcomes are set by faculty and emphasize disciplinary theory, often without connection to the development of skills. Courses are developed independently.	Program learning objectives are validated by stakeholders. An integrated curriculum has mutually supporting courses. Disciplinary fundamentals and essential skills are learned, preparing students for knowledge exchange. [Integrated Curriculum]
The main teaching activities are well-prepared lectures and home assignments. Assessment is usually by individual exam. At its end, the curriculum contains team projects.	Student learning is supported by various evidence-based and authentic learning activities, digital learning tools, and assessment methods. Courses support students in becoming autonomous learners. [Teaching for Learning]
The content of lectures and of the program is well established. Programs have strong disciplinary association and are easily recognized by employers.	Emerging disciplines and bodies of thought flow into the graduate and undergraduate curriculum, despite the fact that these often emerge in the outskirts and intersections of disciplines. [Education in Emerging Thought]
Students can seek further preparation for innovation through optional activities and programs located in business schools or departments.	Students have educational options available to better prepare them for innovation, including programs on technical leadership, management of technology and entrepreneurship. [Preparing for Innovation]

innovation. Other practices relevant to education, including Undergraduate and Postgraduate Student Researchers and Centres of Research, Education, and Innovation, are discussed in Chapter 4.

3.2 Integrated Curriculum

CURRICULUM

3.2.1   Designing a Curriculum for Knowledge Exchange

The foundational practice for education is Integrated Curriculum. It is the process for designing a curriculum guided by a vision of graduates as effective agents of knowledge exchange. The aim is to prepare students not only for knowing, but for *being able to do things that matter* with their understanding. Then the knowledge of disciplinary fundamentals is necessary, but it is not enough.

We call for *deep working knowledge*, implying that students should learn to actively draw on their theoretical understanding when they work on real problems. They also need to develop the *skills, approaches, and judgment* that are essential for a productive working life. These include the personal skills of reasoning, self-awareness, critical thinking, and creative design; and interpersonal skills of communication, emotional intelligence, and collaboration. Approaches to work and life involve ambition and perseverance, as well as respect for others and consideration for sustainable development. Judgment includes the capacity to perceive and analyze situations and handle them appropriately.

To achieve this, a meaningful union is created between theoretical understanding and the essential skills throughout all the courses and projects in the curriculum and in informal learning experiences. Designing the integrated curriculum means creating synergies and making time serve two purposes: acquisition of knowledge and skills.

Box 3.2 Goals of Integrated Curriculum

Universities will educate students with a deep working knowledge
of fundamentals while better preparing them as agents of knowledge
exchange and innovation,
**By implementing an integrated curriculum
of courses, projects, and co-curricular experiences,**
*Educating students who learn the disciplinary fundamentals together with
essential skills, approaches, and judgment.*

The starting point for curriculum design is to formulate the aims of education, informed by a dialog with stakeholders. This vision of what the graduates should be able to do is expressed as intended learning outcomes, related both to the disciplinary knowledge and to the essential skills. The next step is to assign the responsibility for realizing these outcomes to the courses and projects in the curriculum. This is done by apportioning the intended learning outcomes to the subject courses and projects, treating both as sites for developing theoretical understanding together with essential skills. The result is an integrated curriculum consisting of mutually supporting courses and projects, with knowledge and skills integrated throughout (Fig. 3.1). We believe that it is precisely this meaningful relationship that prepares graduates for being able to really do something with their understanding.

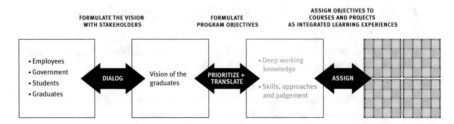

Fig. 3.1 The process of designing the integrated curriculum, guided by a vision of the graduates' desired learning formed in dialog with stakeholders

This approach to curriculum development challenges a traditional view that assumes a zero-sum game between theoretical knowledge and preparation for working life [1]. We do not approach curriculum design as a matter of deciding of "how much" time to spend on each. Instead we seek a meaningful integration of them throughout the curriculum, by emphasizing their interdependence and seeking strong connections between essential skills and the theoretical content of the education [2].

To enrich and broaden the university experience further, many students will also complement the formal curriculum by pursuing various other opportunities for learning. We see the university as a dynamic and diverse learning environment, offering students a broad array of curricular, co-curricular, and extracurricular activities to support their development as whole adults and individuals.

Four key actions for implementing this practice are summarized in Box 3.3, and they are each discussed further below.

Box 3.3 Integrated Curriculum—Practice and Key Actions

Integrated Curriculum is about setting learning objectives informed by stakeholder input, and deploying integrated curricular elements including courses, projects, and co-curricular experiences. This curriculum enables students to build a strong foundation in fundamentals, while simultaneously developing essential professional and life skills, which can be used in innovation. This is the *foundational practice* for education.

Rationale: Implementing a curriculum based on explicit learning outcomes helps ensure that students acquire the appropriate foundations for their future. No doubt, universities are the place where students must learn the disciplinary fundamentals. Integrating essential life and professional skills into courses and projects, as well as rich co-curricular offerings, creates synergies between theoretical fundamentals and skills.

Key actions:
• Engaging with communities of stakeholders—e.g., graduates, employers, government, faculty, and students—to inform the definition of intended learning outcomes.

- Designing mutually supporting disciplinary courses and projects to achieve progression and connections throughout the program.
- Explicitly integrating essential skills with disciplinary fundamentals, so that skills support the learning of fundamentals, and the fundamentals provide the context for developing the skills.
- Encouraging co-curricular activities, including involvement in on-campus research and innovation, and preprofessional and off-campus experiences.

The key outcome is talented graduates.

3.2.2 Engaging Stakeholders to Inform Program Learning Outcomes

The curriculum development process outlined above takes as its starting point a shared vision of the desired attributes of graduates, developed in an ongoing dialog with stakeholders. The result will likely be a partly different and more ambitious guiding vision than in the past.

The first step is to identify the partners who have a legitimate interest in the education and to engage them in discussions. The traditional stakeholder groups for science and engineering education are employers, government, graduates, faculty, and students. A common format for engaging stakeholders is through advisory groups. Other methods, suitable for broader consultation, include seminars, workshops, interviews, and focus groups.

The central question is what knowledge and skills are essential in graduates, which are highly desirable, and which are nice to have. In stakeholder discussions it is important to address priorities and encourage realism, as the risk is otherwise to end up with a long unprioritized wish list of "super powers." It is not possible to do everything for everyone. The internal stakeholders, student representatives and faculty, can have a tempering effect in this respect because they are the ones who will have to realize the vision.

Ultimately, the faculty engaged in the stakeholder dialog must use their judgment to weigh different perspectives against each other and translate the chorus of stakeholder voices into a coherent shape. The effort should finally converge as an ambitious—but still feasible—vision, expressed as intended learning outcomes for the curriculum.

An example of this stakeholder engagement process is presented in Case 3.1: Skoltech—Learning Outcomes Framework. It describes a sequential process of engaging stakeholders and developing learning outcomes. Since the university is built on a mission of applying science and engineering to drive innovation, the learning outcomes reflect key skills around research and creative system and product design. In Chapter 7, a more general discussion about stakeholder engagement is presented.

Case 3.1   Skolkovo Institute of Science and Technology (Skoltech)—Learning Outcomes Framework.

Engaging stakeholders to formulate high-level learning outcomes for a new university.

When Skoltech was started in Moscow, in collaboration with MIT, it embarked on an intensive process to define its educational mission. Through systematic dialog with external stakeholders, the high-level learning outcomes were formulated for the educational programs [1].

The starting point for the process was a 2-day forum in Moscow aimed at identifying the needs of the Russian stakeholders and formulating a corresponding vision of the desired attributes of future Skoltech graduates. The 100 forum participants represented Russian industry, universities and research institutes, international experts on education, and MIT faculty. Synthesizing the discussions, a preliminary list of desired learning outcomes was formulated.

Next, to further develop and validate the learning outcomes, comprehensive interviews were conducted with 38 high-level representatives of large, medium-sized and start-up companies in relevant industry sectors, as well as with research and educational institutions, and governmental organizations. Interpretation and analysis of these interviews indicated both considerable consensus and some tensions.

Engaging in genuine stakeholder dialog does not mean that education should try to do everything for everyone. For Skoltech, such tensions were particularly relevant as it was a university founded for a new mission, whose graduates were expected to drive development and be change agents in industry and enterprise. Stakeholders clearly supported much of what they saw as novel educational ambitions.

The resulting *Skoltech Learning Outcomes Framework for Science Engineering and Innovation Leadership* (see Fig. 3.2) was organized in four sections, corresponding with the UNESCO four pillars of education [2]. It reflects some of the structure of the earlier CDIO Syllabus [3], but has been extended and generalized. Sections 3 and 4 incorporated a four-capability leadership model of relating, sensemaking, visioning and inventing, the latter called "delivering on the vision." [4].

References

1. Edström K, Froumin I, Stanko T, Crawley E (2013) Engaging stakeholders in defining education for innovation in Russia: consensus and tensions. In: Proceedings from the EAIR 35th Annual Forum, The European Higher Education Society. Rotterdam
2. Delors J et al (1996) Learning: the treasure within; Report to UNESCO of the International Commission on Education for the twenty-first century. UNESCO Publishing, Paris, France
3. Crawley E, Malmqvist J, Östlund S et al (2014) Rethinking engineering education: the CDIO Approach, 2nd. Springer, Cham
4. Ancona D, Bresman H (2007) X-teams: how to build teams that lead, innovate and succeed. Harvard Business Review Press, Cambridge, MA

1. DISCIPLINARY KNOWLEDGE AND REASONING

UNESCO PILLAR: LEARNING TO KNOW

1.1 KNOWLEDGE OF MATHEMATICS AND SCIENCES

1.2 KNOWLEDGE OF APPLIED SCIENCE AND ENGINEERING SCIENCE

1.3 KNOWLEDGE OF INNOVATION AND ENTREPRENEURSHIP

1.4 INTERDISCIPLINARY THINKING, KNOWLEDGE STRUCTURE AND INTEGRATION

1.5 KNOWLEDGE AND USE OF CONTEMPORARY METHODS AND TOOLS

2. PERSONAL ATTRIBUTES – THINKING, BELIEFS AND VALUES

UNESCO PILLAR: LEARNING TO BE

2.1 COGNITION AND MODES OF REASONING
- Analytical reasoning and problem solving
- System thinking
- Creative thinking
- Decision making (with ambiguity, urgency etc)
- Critical thinking and meta-cognition

2.2 ATTITUDES AND LEARNING
- Initiative and the willingness to take appropriate risks
- Willingness to make decisions in the face of uncertainty
- Responsibility, intensity, perseverance, urgency and will to deliver
- Resourcefulness, flexibility and an ability to adapt
- Self-awareness and a commitment to self-improvement, lifelong learning and educating

2.3 ETHICS, EQUITY AND OTHER RESPONSIBILITIES
- Ethical action, integrity and courage
- Social responsibility
- Equity and diversity
- Trust and loyalty
- Proactive vision and intention in life

3. RELATING TO OTHERS – COMMUNICATION AND COLLABORATION

UNESCO PILLAR: LEARNING TO WORK WITH OTHERS

3.1 COMMUNICATIONS
- Communications strategy and structure
- Written, electronic and graphical communication
- Oral presentation and discussion
- Inquiry, listening and dialogue

3.2 COMMUNICATIONS IN INTERNATIONAL ENVIRONMENTS
- Communications in English in scientific, business and social settings
- Effective interaction in different cultural and international settings

3.3 TEAMWORK
- Forming effective teams
- Team operations and project management
- Team coordination, decision-making and leadership
- Team growth and evolution
- Technical and multidisciplinary teaming

3.4 COLLABORATION AND CHANGE
- Establishing diverse connections and networking

- Appreciating different roles, perspectives and interests
- Negotiation and conflict resolution
- Advocacy
- Bringing about intentional change

4. LEADING THE INNOVATION PROCESS

UNESCO PILLAR: LEARNING TO DO

4.1 MAKING SENSE OF GLOBAL SOCIETAL, ENVIRONMENTAL AND BUSINESS CONTEXT
- Appreciating the potential and limitations of science and technology, their role in society and society's role in their evolution
- Taking responsibility for sustainable development, including social, economic, environmental and work environment aspects
- Understanding the technical products, systems and infrastructure of the sector
- Understanding the enterprise – culture, stakeholders, strategy and goals
- Understanding the business context – markets, policy and ecosystem of the sector

4.2 VISIONING – INVENTING NEW TECHNOLOGIES THROUGH RESEARCH
- The research process – hypothesis, evidence and defense
- Basic research leading to new scientific discovery
- Research aimed at developing new technologies
- Imagining utility of new science and technology
- Developing concepts and reducing to practice

4.3 VISIONING – CONCEIVING AND DESIGNING SUSTAINABLE SYSTEMS
- Identifying stakeholders need and wants
- Identifying and formulating objectives and goals
- Conceiving and architecting products and services around new technologies and identifying their impact
- Disciplinary and multidisciplinary design for sustainability, safety, aesthetics, operability and other objectives
- Understanding the technical context and ecosystem of the product or service
- Design process management, including planning, project judgment and effective decision-making

4.4 DELIVERING ON THE VISION – IMPLEMENTING AND OPERATING
- Designing and optimizing sustainable and safe implementation and operations
- Manufacturing and supply chain operations
- Supporting the system life cycle including evolution and disposal
- Implementation and operations management

4.5 DELIVERING ON THE VISION – ENTREPRENEURSHIP AND ENTERPRISE
- New venture conceptualization and creation
- Financing product development and new ventures
- Building and leading an organization and extended organization
- Initiating engineering and development processes
- Selling, marketing and distributing products and services
- Understanding the value chain – the innovation system, networks and infrastructure
- Managing intellectual property and respecting the legal process

Fig. 3.2 Skoltech Learning Outcomes Framework for Science Engineering and Innovation Leadership

3.2.3 Designing Courses with Connected Function and Progression

The next stage is to implement the high-level learning outcomes as an educational program consisting of a range of courses and projects. The learning objectives on the curriculum level are broken down and assigned to the various courses and projects. Each curriculum element bears an explicit function and responsibility towards meeting program goals. The learning objectives are then realized in the course or project level design (see Teaching for Learning, below). This approach to curriculum design is solution-independent, as the desired course level learning can be achieved in multiple ways.

The curriculum design process requires intensive dialog across the faculty. The aim is to strengthen *connections* between courses and achieve *progression*, such as when courses depend on each other for prerequisite knowledge and skills, and for appropriate handover when a high-level objective is addressed by a sequence of courses. Effective progression is critical if students are to reach the ambitious learning objectives.

The practice of connecting the courses across the curriculum to address ambitious learning outcomes is well developed at Chalmers University of Technology. Case 3.2 shows how the mechanical engineering faculty are constantly setting new

---

Case 3.2 Chalmers University of Technology (Chalmers)—Integrated Curriculum at Mechanical Engineering

Curriculum is developed to situate learning in relevant contexts and with progression across multiple courses, through a program-driven approach guided by stakeholder input.

Chalmers is a co-founder of the CDIO initiative [1]. The CDIO approach has been applied to continually develop the Mechanical Engineering program. It is a 5-year program combining a Bachelor and Master of Science in Engineering, offering a broad basis in mechanical engineering followed by 15 specializations at the master's degree level. 150 students are admitted annually.

The program development was guided by stakeholder feedback:

- Industry requested stronger design, teamwork, and communication skills.
- The faculty found the mathematics curriculum inadequate in providing modeling and simulation skills.
- Government signaled that engineers must be able to develop sustainable technical solutions.

The resulting *integrated curriculum* is organized around courses typical for mechanical engineering (e.g., mathematics, mechanics, materials, control theory), but with carefully designed *integrated learning experiences* that interconnect different courses or subjects. This ensures progression of learning across multiple courses, exploits opportunities for collaboration between mathematical and technical courses, and anchors students' development of professional skills within the technical context.

One example is the integration of computational mathematics, which strengthened the connection between engineering and mathematics. The rationale was that students need to learn to solve more general, real-world problems, rather than "*solving oversimplified problems that can be expressed analytically and with solutions that are already known in advance*" [2]. Further, students should work on *complete problems*: setting up and solving mathematical models, simulating the system, and using visualization to assess the correctness of model and solution and compare with physical reality. Rather than expecting the mathematics teachers to solve this task within the mathematics courses, a program-driven approach was applied where making connections to mathematics in engineering subjects was at least as important as making connections to engineering in mathematics. Interventions included:

- New basic mathematics courses.
- New teaching materials.
- Integration of relevant mathematics topics in engineering courses.
- Cross-cutting exercises, assignments, and team projects shared between the mechanics and strengths of materials courses and mathematics courses.

Similarly, the integration of sustainable development (see Fig. 3.3) demonstrates how the program-driven approach enables systematic integration of important topics across courses, while maintaining links to program learning outcomes and ensuring progression [3].

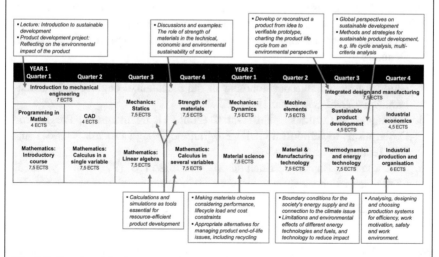

Fig. 3.3 Year 1 and 2 of Mechanical Engineering at Chalmers. Sustainable development is integrated into several courses where sustainability aspects are relevant

The program team implemented the integrated curriculum to keep it unified despite consisting of courses from several departments and disciplines and to further develop the program through an agile process. At Chalmers, programs commission courses from several departments. Every year, program leaders negotiate next year's offering with the delivering department. The agreed course learning objectives, content, pedagogy, and budget are documented in writing. While this is a collegial dialog, the program ultimately controls the budget, approves course syllabi, and receives the course evaluations. The program description [4] helps the program team prioritize among new ideas according to their contribution to the high-level goals of the program [5].

The Mechanical Engineering faculty systematically created conditions for leading, planning, and developing the program, and for constantly setting new goals. The practices developed for the Mechanical Engineering program has had considerable influence across Chalmers, received the highest acclaims in national evaluations and won numerous awards.

Contributed by Professor Johan Malmqvist and Associate Professor Erik Hulthén, Department of Industrial and Materials Science; and Professor Mikael Enelund, Department of Mechanics and Maritime Sciences, Chalmers University of Technology, Gothenburg, Sweden.

References

1. Crawley E, Malmqvist J, Östlund S et al (2014) Rethinking engineering education: the cdio approach, 2nd edn. Springer, Cham
2. Enelund M, Larsson S, Malmqvist J (2011) Integration of computational mathematics education in the mechanical engineering curriculum. In: Proceedings of seventh International CDIO Conference. Copenhagen, Denmark
3. Enelund M, Wedel MK, Lundqvist U, Malmqvist J (2016) Integration of education for sustainable development in the mechanical engineering curriculum. Australas J Eng Educ 19:51–62
4. Malmqvist J, Östlund S, Edström K (2006) Using Integrated Programme Descriptions to Support a CDIO Programme Design Process. World Trans Eng Technol Educ 5:259–262
5. Malmqvist J, Bankel J, Enelund M, Gustafsson G (2010) Ten years of CDIO—experiences from a long-term education development process. In: Proceedings of the sixth International CDIO Conference. Montreal, Canada

goals in dialog with stakeholders and realizing them in an agile process for program-driven course development.

An important condition for forming connections and progression is that educators have an interest in what students will learn in other courses, and in the program as a whole. Another enabler is the participation of faculty role models who have active contact with both research and innovation, who understand the needs of society and industry, and who have an interest in the skills and abilities that matter. These faculty ideals are not new [3], but their full implementation remains a challenge.

3.2.4 Integrating Fundamentals with Essential Life and Professional Skills

The courses and projects of the integrated curriculum should simultaneously address understanding of theoretical fundamentals *and* essential skills, approaches, and judgment. As the building blocks of an integrated curriculum, we call them *integrated learning activities* [2]. Subject courses are designed around some predetermined theoretical content. Here, students will *also* acquire the skills that are relevant for working life, and they will learn them *in the context of the subject* [4]. Learning activities in which students practice skills are also opportunities to enhance their understanding of the theoretical content by actively expressing and applying their new knowledge.

Project courses, on the other hand, are learning activities designed around problems. Here, *in the context of real problems,* students will acquire a *deep working* understanding of theory, and simultaneously develop essential skills, approaches, and judgment. The difference is that exactly what students learn in a project course depends on the needs that arise from the problem at hand. Even the problem statements are not necessarily predefined. Projects can be designed to ensure the learning of some specific theoretical content, or they can be more open-ended allowing the theoretical learning to emerge from the challenges that students encounter.

A key reason to blend problem-led and discipline-led approaches is to support graduates in becoming both problem-oriented and discipline-oriented. By that we mean that they need the sensitivity and curiosity to think in terms of social or economic problems that need solutions, *and* in terms of science and technology with potential for application [5]. Another argument for integrated learning activities is simply to make dual use of time in the curriculum. Practicing skills will also by necessity make the learning experience more active. Bringing in opportunities for using the new knowledge and seeing it from more perspectives also helps deepen their understanding of the disciplinary content. Most importantly, we argue that knowledge and skills give each other meaning.

The TEC21 curriculum at Monterrey Tech, Mexico, aims to shift from mere knowledge acquisition to integrating knowledge and intellectual, technical, and social skills with attitudes and values (Case 3.3). During their studies, students face numerous challenging real-world situations which are increasingly specific and complex. Students also interact with partner organizations.

Case 3.3 Tecnológico de Monterrey—Tec21 Educational Model

Integrating the development of knowledge and intellectual, technical, and social skills with attitudes and values, through challenge-based and flexible learning experiences.

Universities are facing the challenge of preparing students for working in companies not yet created, on problems that may not yet be identified, using technologies not yet invented. Whereas focus in the past was knowledge in itself, we now also value what students can do with what they know, learning to learn, responding appropriately to uncertainty, and adapting to new situations. In this context, Tecnológico de Monterrey announced in 2013 the Tec21 Educational Model, to support students in developing the competencies needed to pursue twenty-first-century opportunities and challenges.

Tecnológico de Monterrey has nearly 10,000 professors, 500 researchers, and over 50,000 undergraduate students on 26 campuses throughout Mexico. The Tec21 initiative began with a comprehensive process of observation, reflection, and design. This included consultation and teamwork with faculty, visits to more than 40 universities worldwide, and an extensive listening exercise with students, graduates, employers, and experts.

Tec21 entails a shift from mere knowledge acquisition to development of competencies, deliberately integrating knowledge and intellectual, technical, and social skills with attitudes and values. The cornerstones are challenged-based learning and flexibility in how, when, and where learning takes place. It provides a memorable university experience with inspiring professors who are up-to-date, connected, innovative, and effectively incorporating technology in the learning process.

In the new undergraduate curriculum, students encounter challenging real-world situations in order to develop their competencies, spark interest and enthusiasm, and stimulate resilience and sense of achievement. Each curriculum unit combines a challenge with learning modules, addressing clearly defined competencies. During the 4 years, a student is exposed to more than 30 increasingly specific and complex challenges. Students are guided and evaluated by an interdisciplinary faculty team and also interact with organizations outside the university.

The 2019 cohort will meet all Tec21 elements in all curricula and at all campuses [1]. During four previous years, the components were gradually tested and implemented with three transition initiatives:

- i-week—involving all undergraduates and faculty members in immersive-challenging activities 1 week/year. Activities engage more than 1000 training partners and include domestic and international trips,
- i-semester—a full semester with six courses following the Tec21 philosophy. To date, more than 5300 students led by more than 1000 professors have developed and executed more than 150 designs,

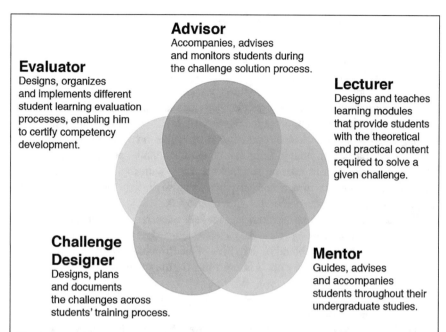

Advisor
Accompanies, advises
and monitors students during
the challenge solution process.

Evaluator
Designs, organizes
and implements different
student learning evaluation
processes, enabling him
to certify competency
development.

Lecturer
Designs and teaches
learning modules
that provide students
with the theoretical
and practical content
required to solve a
given challenge.

**Challenge
Designer**
Designs, plans
and documents
the challenges across
students' training process.

Mentor
Guides, advises
and accompanies
students throughout their
undergraduate studies.

Fig. 3.4 The five roles of educators

- Pathways—flexible degree programs enrolling more than 3000 students.

The incremental development process provided rich feedback for improving the curriculum design. Testimonials of student satisfaction were collected, and educational research demonstrated that the students' learning outcomes are equal to, or better than that of the traditional model. Similarly, companies, civil society organizations, and government agencies are increasingly satisfied with the way the model has improved students' levels of development. Tec21 kindles interest among the international academic community, particularly in Latin America. Also, future students and their parents are responding enthusiastically.

The Tec21 Educational Model is a paradigm shift. The major challenge is faculty development, especially considering that only 22% of faculty members are full professors. The Fig. 3.4 shows the breadth of faculty competence required. Additional academic and operational difficulties emerge during the transition, when old and new programs run in parallel. Nevertheless, faculty and staff are working hard to realize the future of education, demonstrating flexibility, resilience, and openness to challenges.

Contributed by Associate Professor Román Martínez Martínez, Viceprovost for Educational Transformation, Tecnológico de Monterrey, Monterrey, Mexico.

Reference

1. Tecnologico de Monterrey (2019) Tec21 Educational Model Report

3.2.5 Encouraging Co-Curricular Experiences

Co-curricular experiences are less formal but still academically relevant aspect of student life [6]. They take place on campus, in companies, or in the community, and they range from research and innovation projects to preprofessional experiences such as internships. While co-curricular activities are often open-ended, they can still be truly valuable and appropriate as learning experiences. Co-curricular activities have significant intellectual content and offer relevant challenges in authentic contexts. In particular, many activities support students' broader development, creating opportunities to take on responsible roles, with aspects of collaboration, leadership, and organization. Some activities will also directly facilitate student networking and contact with potential future employers.

This holistic view on curriculum and student experience is well understood at Olin College of Engineering (Case 3.4). Here, faculty describe the Olin educational experience as a seamless blend of activities in a curriculum *continuum*, from compulsory courses, to following one's interest just for the fun of it.

Case 3.4 Olin College of Engineering—The Olin Learning Experience

The goal is to educate engineering innovators through authentic projects and interdisciplinary learning, focusing on process outcomes and increasingly independent learning.

Olin College of Engineering is a highly selective undergraduate educational institution with 350 students, graduating its first class in 2006. The mission is to educate *exemplary engineering innovators who recognize needs, design solutions, and engage in creative enterprises for the good of the world*. Olin graduates are seen as some of the world's most successful and innovative individuals [1, 2]. Olin further serves as an educational laboratory, piloting educational experiences, and catalyzing change with partner institutions around the world.

The Olin experience can be described as a seamless blend of courses, projects, co-curricular activities, and doing things just for fun (see Fig. 3.5).

Olin has identified nine *Olin Learning Outcomes*: Application of Knowledge, Skills, Approaches; Critical Thinking; Creativity; Doing Good in the World; Self-Directed Learning; Collaboration; Process Design and Implementation; Communication; and Identity Development. Olin has implemented a curriculum characterized by authentic projects, interdisciplinary learning, process outcomes, and increasingly independent learning [3]:

- From the first semester to the last, students engage in *authentic project-based learning experiences* in which technical skills and knowledge are learned alongside contextual foundations and applications. Student teams select problems, translate them into needs and specifications, then develop prototypes

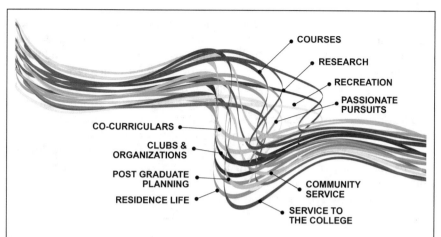

COURSES
RESEARCH
RECREATION
PASSIONATE PURSUITS
CO-CURRICULARS
CLUBS & ORGANIZATIONS
POST GRADUATE PLANNING
COMMUNITY SERVICE
RESIDENCE LIFE
SERVICE TO THE COLLEGE

Fig. 3.5 The Olin College learning continuum

and refine them into products placed in the hands of people. The process involves understanding the human and societal components of technology, and applying human-centered, client-focused design to make something adaptable, adoptable, and sustainable.

- The curriculum contains *interdisciplinary pedagogical approaches aimed at fostering interdisciplinary thinking*. Many courses connect areas of engineering, while also integrating math, science, humanities, and social science. Olin does not have distinct academic departments. Instead, faculty from engineering, mathematics, science, arts, humanities, social science, and entrepreneurship are brought together in a single academic unit. Faculty members are on performance-based renewable multiyear contracts. They are expected to be innovative teachers, whose work is characterized by intellectual vitality and external impact on the academy and the profession.

- Olin emphasizes *process outcomes* rather than comprehensive content coverage. For instance, the Materials Science course covers key areas of materials, but is primarily structured around a project. While students explore deeply *one* particular material, they leave the course ready to apply these processes in new contexts.

- Students are supported to become increasingly *self-directed and autonomous learners*. Faculty often engage students in discussing process—making choices, interpreting outcomes, what might be improved next time—to strengthen metacognition and capability of continued learning, as well as proficiency with the intellectual tools of their discipline. Alumni and employer surveys indicate that Olin graduates are able to work autonomously, even in under-specified environments and on ill-defined problems.

Students in all majors take a common set of core classes. The core includes modeling and simulation, measurement and control, human-centered design, and an introductory entrepreneurship course, where students start developing an entrepreneurial mindset and learn tools essential for realizing sustainable change. Later semesters offer more opportunities for diverging paths, while also intermingling students across levels and majors. The culmination is a student-directed final year capstone. Many students also undertake a one-semester culminating experience in arts, humanities, social sciences, or in entrepreneurship. Mentored by skilled practitioners, students *actualize a vision*, producing an actual or virtual artifact that is of recognizable value to disciplinary experts.

Contributed by Professor Lynn Andrea Stein and Professor Rob Martello, Olin College of Engineering, Needham, Massachusetts, USA.

References

1. Edwards L (2017) The college that produces founders at 5 times the rate of Stanford. Medium, blogpost. In: blog.ledwards.com. http://blog.ledwards.com/the-college-that-produces-founders-at-3-times-the-rate-of-stanford-2c53ea44f91e. Accessed 20 Jan 2020
2. Graham R (2018) Global state of the art in engineering education. Massachusetts Institute of Technology, Cambridge, MA
3. Stein LA, Somerville MH, Townsend J, Manno V (2013) Olin college: re-visioning undergraduate engineering education. In: DeVitis JL (ed.) The college curriculum: a reader. Peter Lang, New York, pp. 249–265

In addition to playing an important role as a part of education, co-curricular activities can express the innovative spirit of the university. Students seek such opportunities with enthusiasm and creativity, and the activities may sometimes have a playful, rebellious, and subversive flavor. While these activities are outside classroom settings, the guidance and support of faculty mentors can make a big difference. A lighter touch is recommended, however, as co-curricular experiences might lose some of their motivational appeal if they were organized by faculty. They are still opportunities for faculty and students to meet informally and share the sheer joy of serious play.

3.3   Teaching for Learning

LEARNING

3.3.1   Learning for New Challenges

It is reasonable to assume that when the aims of education have been identified in dialog with stakeholders, they imply high expectations on student learning. These desired learning outcomes will be a combination of knowledge, skills, approaches, and judgment, needed for a working life of knowledge exchange.

Teaching for Learning is the practice of engaging students in active, experiential, and digital learning activities. Guided by the practice of Integrated Curriculum, the intended learning outcomes on the program level were assigned to curricular elements: courses and projects. The challenge is now to design courses and projects to address the desired learning. We focus on how courses can support students in developing deep working understanding of disciplinary fundamentals, while also furthering self-efficacy and preparing for a life of independent learning.

Box 3.4 Goals of Teaching for Learning

Universities will educate students with a deep working knowledge
of fundamentals while better preparing them as agents of knowledge
exchange and innovation,
By engaging students in active, experiential and digital learning,
*Graduating students with deep working understanding, self-efficacy, and
a capability for self-learning.*

Table 3.3 Feisel-Schmitz technical taxonomy

Judge	To be able to critically evaluate multiple solutions and select an optimum solution
Solve	To be able to characterize, analyze, and synthesize to model a system (make appropriate assumptions)
Explain	To be able to state the process/outcome/concept in their own words
Compute	To be able to follow rules and procedures (substitute quantities correctly into equations and arrive at a correct result, "plug & chug")
Define	To be able to state the definition of the concept or describe in a qualitative or quantitative manner

When we refer to *deep working understanding*, we imply that courses should be designed to aim for high-quality learning. The nature of quality in learning is illustrated by the Feisel-Schmitz technical taxonomy [7], which captures the distinction between applying knowledge *with* or *without* understanding.

The taxonomy is nested, meaning that each level subsumes those beneath, and therefore it is read from below (Table 3.3). On the lower levels, *Define* and *Compute* correspond to being able to repeat course content and pattern-matching typical given problems with given solutions. Learning that is limited to these levels is not sufficient to prepare students for work and life.

The *Explain* level denotes the ability to articulate concepts with meaningful understanding. Performances on this level include solving familiar problems while *also* understanding what they did and why, and being able to interpret the results. The *Solve* level denotes the ability to make connections within the material and with previous learning. As students create a coherent and meaningful knowledge structure, they can take on also such problems that are new to them. Finally, *Judge* is the ability to inform decisions with analysis. Clearly, these higher levels of understanding are required to lay the foundation for a productive working life, and for contributing to knowledge exchange.

In the following, we describe several principles and strategies for designing courses. Different forms of active learning and meaningful assessment will together create a motivational context in which these high expectations on learning can be realized.

Four key actions are recommended to achieve high quality of learning, as indicated in Box 3.5.

Box 3.5 Teaching for Learning—Practice and Key Actions

Teaching for Learning is about deploying teaching and assessment approaches which align with intended learning outcomes, engage students' attention and curiosity, and involve active, experiential, and digital forms of learning, in order to further deeper working understanding, confidence and self-efficacy, and capabilities for reflection and self-learning.

Rationale: When students are more actively engaged in their learning and allowed to express and try out their knowledge in practical applications, their

understanding becomes deeper and more active. This allows them to manipu-
late knowledge in ways that supports the creation of new inventions and products. This process is also enabled by self-efficacy and self-learning.

Key actions:
- Constructive alignment of intended learning outcomes, learning activities
  that support learning of skills and fundamentals, and assessment
  activities.
- Active learning, engaging students in manipulating and evaluating ideas,
  and experiential learning in situations resembling working life, leading to
  the development of self-efficacy.
- Digital learning—so that students can access many resources and points of
  view, blended with face-to-face learning.
- Encouraging self-learning—the ability to reflect on past experiences, identify and satisfy the individual's need for new knowledge and skills.

The key outcome is talented graduates.

3.3.2   Constructively Aligning Intended Learning Outcomes, Activities, and Assessment

Constructive alignment is a model for course design, which holds that learning
activities and assessment systems should be designed in purposeful relation to the
intended learning outcomes [8]. The learning activities should engage students in
appropriate work, through which they can achieve the desired learning. The assessment system should judge student performance to assure that they have learnt what
was intended (Fig. 3.6).

The aim of constructive alignment is to promote a *deep approach to learning*,
indicated by students' intention to understand and make sense, and strongly correlated with well-structured and persistent learning [9]. The deep approach is encouraged when there is a motivational context for learning, when learning is active and
requires interaction with others, and when assessment tasks demand meaningful
performances that require deep understanding. Different forms of such learning are
discussed below. The opposite to a deep approach is a surface approach, meaning
that students just intend to skim along and pass the course. Since the surface
approach is associated with poorly structured learning with little chance of reten-
tion, it should be discouraged [8].

The question guiding the design of learning activities is: "What work is appropriate for the students to do, to reach the learning outcomes?" Learning activities are
designed to capture *sufficient* student time and effort, and more importantly, to generate *appropriate activities* that can result in quality learning and ownership of the
knowledge.

Fig. 3.6 Constructive
alignment of learning
outcomes, learning
activities and assessment,
and the associated
questions to guide course
design

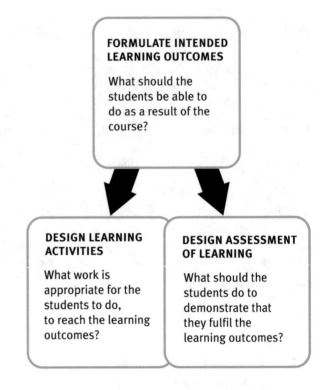

It also matters how assessment systems are designed and communicated, as they have powerful effects on student behavior and learning. For one thing, assessment can influence whether students adopt a surface or deep approach. Assessment will also direct their attention to what is important, act as an incentive for study, affect what they do and when they do it, challenge them, and build their confidence for future work [10].

3.3.3   Stimulating Active Learning

Active learning approaches cover a range of methods including active engagement and experiential learning [11]. Students learn by engaging with new ideas, expressing them and trying them out in action, often in interaction with peers and experts. There is evidence that active learning is conducive to learning [12].

The value of active student engagement with the disciplinary concepts is well demonstrated in Eric Mazur's teaching at Harvard University (Case 3.5). He developed Peer Instruction [13], an early and powerful example of what is now often referred to as "flipped classroom" methods. The basic idea is that students read material in advance, and class time is then devoted to higher level activities, such as retrieving, expressing, and applying what they have learned, in discussion with peers and teachers. In Eric Mazur's class, time is used for strengthening student understanding and addressing misconceptions, through active retrieval as a response to concept questions and discussion with peers. Retrieving knowledge has been shown to have strong positive effects on learning, making it more effective than repeated teaching [14].

Case 3.5 Harvard University—Peer Instruction

Improving conceptual understanding and problem-solving skills through a teaching format in which students learn by actively and interactively exploring their understanding.

I teach introductory physics for undergraduates at Harvard University. During my first years as a teacher, my course was successful by all measures. Students could solve difficult problems, and they rated my lectures highly. Then, in 1990, I discovered that, despite being satisfied and passing exams, my students did not internalize the basics. A test designed by Halloun and Hestenes [1, 2] showed that my students held many of the typical misconceptions relating to core concepts such as force and motion. While they could recite the laws and apply them in advanced numerical problems, my students did not really understand them in any deeper sense. It became obvious that they had concentrated on memorizing procedures for solving typical problems rather than understanding [3].

To address the need for deeper conceptual understanding without sacrificing students' ability to solve problems, I developed a new way of teaching called *Peer Instruction* [4]. Essentially, instead of lecturing, I am now teaching by asking questions that are designed to address challenging concepts and reveal commonly held misconceptions.

Peer instruction is structured around conceptual questions, usually (but not necessarily) with multiple-choice answers. Prior to class, students read the assigned materials. During class, we follow a procedure for probing and developing students' understanding by working together on a series of questions ("ConcepTests") (see Fig. 3.7).

- After thinking about the first question individually for a few minutes, students are polled for their answers using apps, handheld devices, or colored flash cards.
- If 30–70% of the students answer correctly, I ask them to discuss with each other and try to convince someone who had a different answer. My teaching team circulates the room to promote discussions and encourage productive arguments. After a few minutes, students submit their revised answer. Normally, the proportion of correct answers increases substantially as a result of the discussion.

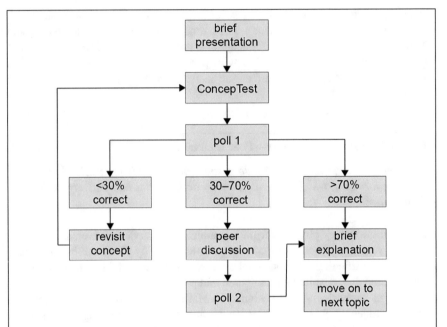

Fig. 3.7 Peer Instruction flowchart

- If more than 70% of the students have the right answer, I just give a very brief explanation and move on to the next topic.
- If fewer than 30% of the students get it right, it does not make sense for them to discuss, as there are too few students to convince others of the right answer. In that case, I give an explanation and repeat the process with a different concept question on the same topic.

With peer instruction, learning gains have increased significantly. Students perform better on conceptual assessments, but, importantly, their traditional problem-solving skills are also improved. This has been the experience, not only in my physics classroom [5], but also for colleagues worldwide in a range of subjects and at numerous institutions [6, 7].

Peer instruction engages students in thinking for themselves and expressing their understanding in words. It provides frequent and continuous feedback to the students and to the teacher. It forces students to expose and confront their understanding and its limitations, in a setting where they have the opportunity to resolve misunderstandings and work together with peers and teaching staff to learn the ideas and skills of a discipline.

Contributed by Eric Mazur, Balkanski Professor of Physics and Applied Physics and Area Chair for Applied Physics, Harvard School of Engineering and Applied Sciences, Cambridge, Massachusetts, USA.

References

1. Halloun IA, Hestenes D (1985) The initial knowledge state of college physics students. Am J Phys 53:1043–1055
2. Halloun IA, Hestenes D (1985) Common sense concepts about motion. Am J Phys 53:1056–1065
3. Hestenes D, Wells M, Swackhamer G (1992) Force concept inventory. Phys Teach 30:141–158
4. Mazur E (1997) Peer instruction: a user's manual. Prentice Hall, Upper Saddle River, NJ
5. Crouch C, Mazur E (2001) Peer instruction: ten years of experience and results. Am J Phys 69:970–977
6. Fagen AP, Crouch C, Mazur E (2002) Peer instruction: results from a range of classrooms. Phys Teach 40:206–209
7. Lasry N, Mazur E, Watkins J (2008) Peer instruction: from Harvard to the two-year college. Am J Phys 76:1066–1069

Given the aim to prepare students for taking on consequential work, a key concept is self-efficacy [15]. This is the *belief in one's capability* to successfully produce a desired outcome in a given situation. It affects students' performance, thinking, motivation, and behavior, and there is strong correlation between self-efficacy beliefs and actual success. People with high self-efficacy set themselves challenging goals and maintain stronger commitment, and handle failure more constructively. Active learning involves making experiences and interacting with others, which can help build self-efficacy.

When the development of essential skills is integrated into theoretical courses, the learning experiences will become more active. Skills are best developed through repeatedly practicing and performing *in the context of the subject* and aided by reflection and by interaction including feedback. This goes for personal skills—such as critical thinking, perseverance, and responsibility, as well as interpersonal skills—such as communication, collaboration, and negotiation.

In *experiential* learning students participate in *authentic* experience, such as design projects or case studies where students assume roles like those that they might encounter in working life. Just "learning by doing" is not sufficient, however; it is only through reflection on the experience that it is transformed into learning. Here, faculty have key roles in providing students with opportunities, structures, and guidance for interpreting their own experience, using feedback from others productively, and developing confidence in their own judgment.

The practice of guiding the student experience in project- and problem-based education is well developed at Aalborg university, which is based on this pedagogy (Case 3.6). The Aalborg curriculum is dominated by projects, and students are guided in reflection by faculty, to develop process competencies.

Case 3.6 Aalborg University—Problem- and Project-Based Learning

A curriculum rich with projects where student teams work on authentic problems to develop professionally relevant competencies through guided reflection.

Aalborg University was established in 1974 as a reform university based on a new pedagogy, today called *problem- and project-based learning* (PBL). In the Aalborg PBL model, during each semester students share their time equally between projects and supporting disciplinary courses. Students work in teams of 5–8, on problems they define themselves within a thematic framework, or on authentic problems from society. In the case of engineering education, problems are often derived from companies. Project teams interact with the companies, seeking information, or forming various degrees of collaboration. The organization of the projects differs. While during one semester, students participate in three small projects, during another, they form mega-projects with several project teams working together [1].

In parallel with the project work, students are guided by academic staff to reflect on their *process competencies*, most notably the ability to communicate and collaborate internally and externally with facilitators, stakeholders, and project management. Other skills include conflict management, information seeking, etc. Working on projects is both a way to organize the learning process; and to develop such competences that are necessary in working life and best learned through experience. Further, the more problems resemble real life, the more students experience a sense of meaning and motivation to learn.

Aalborg University has expanded vigorously, with three campuses, five faculties, and more than 22,000 students. It is frequently ranked by Danish companies as the best university to collaborate with [2]. A study of engineering students in their final semester found that, compared to students from other Danish institutions, Aalborg University students see themselves as equally well prepared in terms of technical knowledge and better prepared for all other aspects of working life (e.g., societal context, contemporary issues, social responsibility, environmental impact, communication, teamwork, management skills, and problem-solving) [3].

Aalborg University has a new digital strategy, *Knowledge for the World*, with profound implications for its PBL model. The aim is to equip students to develop the digital skills required in society in which digitalization plays an ever-increasing role. Students need to reflect critically on the implications of digitalization in relation to their discipline. In their studies, they should use and reflect on the digital opportunities available for communication, cooperation, and information search processes. The taught courses will increasingly adopt digital modes, offerings opportunities to learn just-in-time, as the knowledge is needed in their projects. Hence, the projects will become even more important as a way to drive student learning of disciplinary knowledge.

For many years, Aalborg University has served as a living lab, with international visitors observing and interacting with students and academic staff. In the

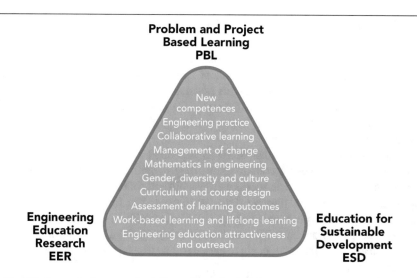

Fig. 3.8 Research themes at the Aalborg Centre For Problem-Based Learning in Engineering Science and Sustainability under the Auspices of UNESCO (http://ucpbl.net)

engineering education community, Aalborg University was internationally recognized as the fourth most influencial university [4]. The UNESCO PBL Centre at Aalborg conducts research on PBL and sustainable development in engineering education (see Fig. 3.8). The Aalborg Centre is a core actor in global networks and provides training and education for academic staff and institutions wanting to adopt PBL.

Contributed by Professor Anette Kolmos, UNESCO Chair in Problem-Based Learning in Engineering Education, Aalborg University, Aalborg, Denmark.

References

1. Kolmos A, de Graaf E (2014) Problem-based and project-based learning in engineering education: merging models. In: Johri A, Olds BM (eds) Cambridge handbook of engineering education research. Cambridge University Press, New York, pp. 141–161
2. Hoff A, Orebo Pyndt J, Rønhof C (2018) Danmark tilbage på vidensporet (Denmark Back on the Knowledge Track. The Confederation of Danish Industry, Copenhagen, Denmark
3. Kolmos A, Holgaard JE (2017) Impact of PBL and company interaction on the transition from engineering education to work. In: sixth International Research Symposium on PBL: PBL, Social Progress and Sustainability. Aalborg Universitetsforlag, pp. 87–89
4. Graham R (2018) Global state of the art in engineering education. Massachusetts Institute of Technology, Cambridge, MA

3.3.4   Digital and Blended Learning

Students take advantage of digital access and tools during their education, both in ways that are planned by their instructors and using their own, sometimes more agile, digital skills. The rationale of this practice is to increase the scope of learning utilizing digital resources and tools, to enhance the effectiveness of education, to increase students' reach in their future work, and to support a life of continuous independent learning.

Long gone are the days when students depended only on instructors and textbooks for access to course content. Now, they make their own choices and also control the pace. Students can access well-established content and emerging thought from all over the world, through digital resources. The information comes from various sources, from peer-reviewed scientific information to everything in the commercial and public domain. We recommend using digital access and tools not to replace face-to-face interaction, but to enrich it. The flipped classroom, for example, does that.

In projects and in informal study groups, students practice working modes where communication and collaboration are enhanced by digital and blended formats. The practice of using a blended format in projects is well developed at Tsinghua University (Case 3.7). Following a structured approach called Extreme Learning Process, student teams take on challenges while all the time capturing and sharing their work in a digital environment.

Case 3.7 Tsinghua University—Extreme Learning Process

Enabling scalable challenge-based learning through standardized educational processes supported by a digital learning environment.

The Extreme Learning Process (XLP) is a challenge-based collaborative learning process for educating digital natives for our digital world. It allows efficient scalable delivery of learning to large groups of learners from different disciplines, at any level of education or professional experience. It does this by integrating the latest educational processes and knowledge management technologies.

In an XLP course, instructors guide student teams through the challenges, following certain educational processes:

- **Early Success**—Participants first successfully execute a simple version of the challenge, often through novel application of digital tools, providing inspiration and faith that it can be done.
- **Safe Failures**—Participants are encouraged to experiment and learn, sometimes failing gracefully and always learning quickly.
- **Convergence**—After failing safely, individuals learn that success comes from teamwork. From collective experience, multifunctional teams work towards shared demonstrable results.

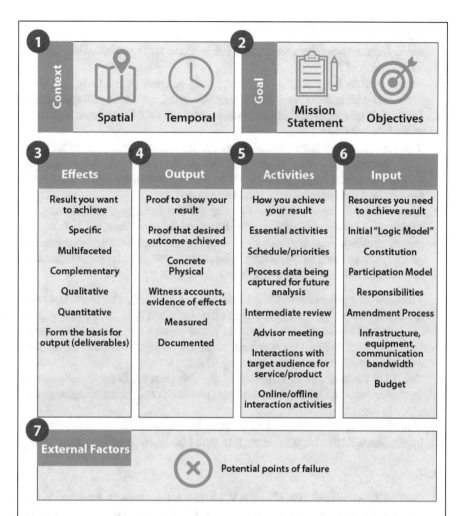

Fig. 3.9 XLP's Logic Model with a sequence of seven items to be completed by learners

- **Demonstration**—As one team, the group gives a powerful demonstration of what has been achieved. This demonstration, the learning it demonstrates, and the confidence it provides is the true reward.

XLP learners work together within and between teams, immersed in a digital environment using standardized processes. This allows previous participants to immediately share their work to help teach others. This enables large-scale deployment and the education of digital natives by their native peers through self-governance.

Learning is supported by the XLP digital platform, using the newest knowledge management technologies, based on four tools:

- **The logic model**—As a pathway to student achievement, each team structures their work following the sequence shown in the Fig. 3.9. They document their work with digital tools, providing time stamp and version control features. Learner behavior can then be tracked and analyzed, allowing insight into patterns of learning.

- **Smart contracts**—Using a digital template, learners define their contributions and validate contract fulfillment. Contract execution infrastructures keep track of specific privileges and responsibilities of each team member, using blockchain-inspired technology.
- **The constitution**—Each student group customizes a reference version of the XLP constitution, a set of high-level rules binding every XLP participant. It outlines tasks and responsibilities, informs students of their rights, and benefits and lays out approaches to resolve conflicts.
- **Digital-publishing workflow**—Participants use industry-standard tools to package, archive and digitally publish the outcomes of their work as documents, similar to industry analysis reports or product plans. The present technology includes data navigation and visualization tools, with plans to also add data extraction, analysis, and machine learning.

Professional learners have also used XLP successfully [1]. Professional teams have published real consulting reports in relatively short time. Small enterprises have completed business plans in 1 week. These teams were often multinational and multicultural. While they were not made up of digital natives, the digital tools were learned on the fly with tutoring.

The XLP method was first applied in the Tsinghua challenge-based course series. Since June 2012, XLP-based courses have been conducted at various universities in Asia.

Contributed by Dr. Benjamin Hsueh-Yung Koo, founder of XLP and Director of International Relations, iCenter of Tsinghua University, Beijing, China.

Reference

1. Tsai WT, Alimbekov K, Koo BHY (2015) An evaluation framework for extreme learning process (XLP). In: 2015 IEEE Symposium on Service-Oriented System. IEEE, pp. 357–366

Students should also have access to professional sources for information retrieval as well as tools for computation, simulation, and visualization. Another opportunity created by digitalization is to guide students in developing a scholarly or professional identity and connect to relevant communities, by actively cultivating a personal learning network [16]. Digital channels offer access to peers, professionals, thought leaders, and role models, and to current issues and debates. As students become more established over time, they can also increase their own visibility and access opportunities through these networks.

3.3.5 Learning to Learn

In working life, graduates will continuously face new kinds of challenges. This happens because they take on new roles, because new knowledge is constantly evolving, and because conditions, organizations, and societies change. It can therefore be argued that the most important purpose of education is to equip students for independent learning for the rest of their lives, without the supportive structure of educators, peers, schedules, and examinations. When devising learning activities and assessments, it is helpful to consider not only how to support the present learning objectives, but also how the experiences can help students develop their ability to monitor and control their own learning [10, 17].

In contrast to the somewhat artificial context of traditional university courses, we believe that when learning and assessment tasks resemble those in working life, students will establish useful self-learning habits which are more easily invoked in future learning situations.

To develop as independent learners through authentic assignments and projects, students need opportunities to strengthen their habits of self-reflection. Structured approaches to stimulate reflection include self-evaluation exercises and "lessons learned" writing tasks.

Feedback is key to provoke deeper reflection and help students avoid "cooking in their own juice." Feedback can come from peers, educators, other experts, or the users of student work. Feedback that directs attention to the learning process can further strengthen students' ability to evaluate and manage their own learning [18].

3.4 Education in Emerging and Cross-Disciplinary Thought

3.4.1   Education at the Boundaries of Knowledge

Classical knowledge is timeless. But now the growth of scholarly knowledge is accelerating, traditional disciplinary boundaries are blurring, and new disciplines are emerging. This emerging and cross-disciplinary thought is important in knowledge exchange, and often forms the basis for new research and innovation. We owe it to our students to prepare them in relevant emerging bodies of thought. *Education in Emerging Thought* is a strategy for furthering effective knowledge exchange through education.

Box 3.6   Goals of Education in Emerging Thought

Universities will educate students with a deep working knowledge of fundamentals while better preparing them as agents of knowledge exchange and innovation,

By migrating cross-disciplinary and emerging research outcomes quickly to the curriculum and learning opportunities

Preparing students who learn emerging disciplines, technologies, and bodies of thought.

Within any given discipline, the pace of emerging thought is increasing. The number of research-intensive universities has multiplied in a few decades. More scholars, better facilities, better prepared students, faster communication, and more publications all contribute to the increased pace. This is particularly true in the sciences and technology, where science accelerates technology, and new technology plays a part in enabling new science.

Cross-disciplinary research also intensifies the development of emerging understandings, and the codification of the new knowledge often spans disciplinary and institutional boundaries. Such efforts, supported by funding bodies and university strategies, contribute new, rich bodies of knowledge.

Preparing students in relevant emerging and cross-disciplinary fields will make them attractive agents of knowledge exchange. Emerging bodies of thought are of special interest for knowledge exchange precisely because they are new, potentially enabling innovation and entrepreneurship (Chapter 5). Exposure to emerging thought also makes students better researchers and more likely to gain an academic position (Chapter 4). If university graduates are to benefit from these opportunities, they must be able to access emerging knowledge through the curriculum and other learning activities.

Innovation and scientific discovery also arise from a recognition of new opportunities and capabilities in the scientific, technical, or social context. When a truly new discovery is made—something of the importance of the LASER or DNA—it is an opportunity to stop and consider the implications for education. The same needs to be done when a new social need is recognized or a new scientific facility comes online. Instruction in emerging thought should aim to support students in becoming more aware of such new opportunities.

Three key actions recommended to support education in emerging thought are given in Box 3.7.

Box 3.7 Education in Emerging Thought—Practice and Key Actions

Emerging Thought is about migrating knowledge from research and deploying it in learning activities and curriculum that support student learning of emerging fields, disciplines, technologies or bodies of thought, or that are based on new opportunities and capabilities, and which might cross existing disciplinary or institutional boundaries.

Rationale: Breakthroughs in research and innovation are frequently made at the boundaries between disciplines, in areas of new emerging thought, or because of new opportunities in the scientific, technical, or social context. Education in emerging and cross-disciplinary topics equips students with contemporary knowledge, strengthening their contributions to research, and accelerating innovation.

Key actions:
- Prompt transfer into curricular elements and online learning of emerging ideas and the awareness of new opportunities.
- Developing learning opportunities in fields transcending existing departmental or disciplinary structures, promoting interdisciplinary and transdisciplinary learning.
- Recognizing the newly developed curricular elements and programs as part of the institutional structure and providing sustainable resources.

The key outcome is talented graduates although there is close linkage to research discoveries that form the basis for this practice.

3.4.2   Promptly Transferring New Discoveries to the Education

Since we cannot teach our students everything, it is a matter of constant reappraisal which topics might be most impactful in the curriculum. To prepare students as agents of knowledge exchange, they need to learn fundamentals established long ago as well as contemporary bodies of thought based on more recent discoveries. This will not only benefit the graduates in their working life, but also the laboratories, companies, and organizations that they engage in, and ultimately society.

Universities have a tradition of stability, influenced by regulations, accreditation bodies or national authorities, and the nature of stable funding streams. Nevertheless, stable curricula are in tension with the need to accelerate innovation in society through knowledge exchange of emerging and cross-disciplinary thought.

It is appropriate to adopt a more agile approach to curricula. Universities can form different pathways for students to encounter new discoveries promptly, even within weeks or months. Using a more conventional approach, emerging ideas can be directly transferred as new modules into postgraduate courses. These new modules can grow in time to become new courses and even programs. Eventually, this emerging thought can migrate to the undergraduate curriculum. New thought can also be introduced into education through undergraduate and postgraduate students reading and writing journal articles and attending conferences.

Project-based learning in postgraduate classes provides a flexible mechanism for exploring relevant emergent topics. New knowledge is brought in as needed into the projects, which can also be about investigating new opportunities.

One new pathway to emerging thought is digital. Quick access is available through: inspirational talks that flag up new opportunities; online video lectures and courses at the frontier of knowledge; and open courseware (Case 3.8). Online journals, scholarly repositories and patent databases connect new discoveries to all. In particular, digital education platforms, like edX, provides the ability to rapidly disseminate knowledge at the leading edge of research, making it much more widely accessible than otherwise possible.

Case 3.8 edX—Acquiring Emerging Thought Through Online Learning

In leading university programs, new research outcomes move quickly to the curriculum through a set of conventional and online approaches.

Conventionally, new research results influences education when a researcher writes a paper that is read by other faculty members worldwide. Depending on the independence of faculty members as instructors, they may introduce the new finding in the curriculum, probably first in specialized graduate courses, then in more foundational graduate courses and, eventually, undergraduate courses. For instance, much of what we teach in molecular biology was discovered in the last 60 years.

At research universities, individual researchers sometimes make a discovery that is quickly introduced into teaching directly by the researcher. In fact, one of the comparative advantages of major research universities is the ability to stimulate this kind of rapid transfer from research to teaching.

This comparative advantage may disappear in the era of online education. Among the functions that digital education platforms can provide is the ability to post learning modules at the leading edge of research online, quickly, and for all to access and use.

At edX [1] and its peer programs, there may not explicitly be a category for such leading edge offerings, but they are clearly there. For example, the edX courses "Principles of Synthetic Biology" and "Photonic Integrated Circuits" are at the frontiers of knowledge (see Figs. 3.10 and 3.11). The advantages of these online offerings are that they are easily available, coherently presented, well produced and taught by an expert.

At an institutional level, some universities use offerings to strategic advantage. For example, Chalmers leads the flagship EU research program on Graphene. To distribute its research outcomes to the educational market, Chalmers maintains an edX Massive Open Online Course (MOOC) on Graphene.

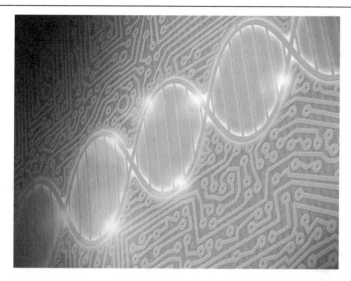

Fig. 3.10 Illustration from the edX MOOCs on Principles of Synthetic Biology (Photo courtesy Ron Weiss, MIT)

Fig. 3.11 Illustration from the edX MOOCs on Photonic Integrated Circuits: AIM Photonics education teaching PIC chip, designed by RIT Integrated Photonics Group and fabricated by AIM Photonics Institute (Photo courtesy Michael Fanto, RIT)

There are several disadvantages of relying on digital education for this kind of leading edge learning:

- Market pressure will tend to favor production of lower level "introduction to …" or "fundamentals of …" offerings that are accessible to a wider audience.

- The effort to plan, produce, and deliver such an offering will introduce a lag up to a year.
- Because of the effort required for updating, offerings may become static, and out of date after several years.

MOOC quality, relevance, and timeliness can be ensured with strong and steady feedback from faculty experts. In addition, there is competition among the offerings by different universities on edX, and between edX and its peer programs. Both of these mechanisms work to keep the offerings fresh.

The result of postings of this type to edX is to make broadly available material created experts at the leading edge of knowledge. The online interaction can provide a kind of crowdsourcing to assure the quality and accuracy of the material, suggest potential revisions, and provide indications of when material should be taken down.

Prepared with the assistance of Professor Sanjay Sarma, MIT Vice President for Open Learning, Massachusetts Institute of Technology, Cambridge, Massachusetts, USA.

Reference

1. edX. https://www.edx.org. Accessed 20 Jan 2020

3.4.3   Education in Cross-Disciplinary Fields

In addition to learning emerging knowledge, we want to prepare students to be able to work across disciplinary fields. The nature of such cross-disciplinary activities spans a spectrum. In multidisciplinary research, scholars from different fields address common questions, each *adding* the respective approaches. Interdisciplinary research fields have a more interactive character, seeking *agreement* in the methodological approaches. Transdisciplinary research deals with issues where the existing domains are not sufficient, and therefore the resulting approaches may end up *outside the map of existing disciplines* [19] (See Chapter 4).

Education in emerging cross-disciplinary fields prepares students to address challenges that do not fit comfortably within a single discipline. It also makes them ready for working and communicating in cross-disciplinary environments and collaborations. They need to embrace an attitude of intellectual openness, learning to handle different perspectives while exploring new ideas in respectful dialog. The challenge for faculty is to consolidate the knowledge in forms appropriate for university teaching, and to role model the constructive attitudes.

The more conventional approach to cross-disciplinary learning is to create new educational structures that are cross-disciplinary. These could be programs, such as a graduate program on computational engineering. It could lead to a new department, such as a department of genetics or mechatronics. These tend to redraw the curricular boundaries and are a more stable cross-disciplinary structure.

Another option is to assign postgraduate educational responsibility to an inter-disciplinary research center. For example, a research program on health science and technology could offer a doctorate of the same theme. These interdisciplinary research centers can also take on innovation activities, creating an integrated Centre of Research, Education, and Innovation (Chapter 4).

An alternative approach is to assign the responsibility for cross-disciplinary linking to the postgraduate students. In this model, the students are free to enroll in courses and have advisors in two otherwise separate programs. The students participate in the seminars and intellectual life of both sides. They create a bridging function that can also benefit the organizations.

An example of such a program is at the Nanyang Technological University in Singapore. It offers multidisciplinary PhD education programs. The students participating in these programs contribute directly to cultivating an interdisciplinary culture at the university (Case 3.9).

Case 3.9 Nanyang Technological University (NTU)—Interdisciplinary Degree Programs

Students can be empowered as agents of cross-disciplinary knowledge creation and sharing by creating interdisciplinary degree programs as a complement to disciplinary programs in schools and departments.

In today's increasingly complex world, societal problems are often multifaceted. To address such challenges, scholars and students must possess cross-disciplinary knowledge. A strength of NTU is its academic programs that cross disciplines, among scientific and technical fields, as well as encompassing social science, arts, humanities, business, and management. The students in these programs can serve as agents for the flow of interdisciplinary knowledge.

To achieve this goal, these programs offer a multidisciplinary curriculum, guidance by at least two top faculty, intensive research opportunities and overseas exposure, as well as dialog with world-class scientists and industry leaders.

At the doctoral level, the Interdisciplinary Graduate Program (IGP) both educates students and drives the development of a culture of interdisciplinarity campus wide [1]. It enables students to be co-creators of knowledge and produces a push in PhD education that will erode old barriers and drive research into areas that benefit from a multidisciplinary approach. Notably, NTU is creating an academic program to do this, not a new research program or center. The students will be the main agents of the interdisciplinary effort.

To enroll in the IGP, students submit an application to the newly established Graduate College [2]. IGP provides students a home for interdisciplinary encounters, discussions, and courses while the students maintain full affiliation with the schools of their supervisors, where they will regularly attend seminars and participate in academic life. The regulations allow these PhD students to complete two out of their six graduate courses in a second discipline, as compared to regular PhD students who take all six courses in their core discipline. Students may also choose to complete 1 year of academic research in top universities overseas.

PhD students undertake a multidisciplinary approach attuned to globally complex issues, currently in one of five key areas: Sustainable Earth, Healthy Society, Secure Community, Global Asia, and Future Learning, tapping into the faculty and talents in the university's research centers of excellence, research institutes, colleges, and schools.

For example, the Healthy Society specialization provides support for a new generation of interdisciplinary professionals to meet the healthcare needs of Singapore and beyond, taking advantage of NTU's Lee Kong Chian School of Medicine, set up jointly with Imperial College London. The Secure Community specialization promotes collaboration between the engineering and social sciences, bridges media systems and content research, and challenges the boundaries of innovation in digital media in areas such as homeland and cyber security, artificial intelligence, virtual reality, crowd simulation, 3D fashion, social robotics, and IT for the elderly.

To exchange knowledge internationally, NTU runs joint and dual PhD degree programs with key global partners, such as Imperial College, London; the Technical University of Munich; Sorbonne Université, Paris; and the University of California, Berkeley. NTU also has struck connections with global industry partners such as Alibaba, BMW Group, and Rolls-Royce.

The university stimulates a multidisciplinary culture through the award of grants for multidisciplinary research. The outcomes of these efforts flow first to research students, and then into courses, at the initiative of individual faculty members. The university also helps by providing innovative teaching grants and awards for accelerating creativity.

Prepared based on input provided by Alan Chan, Vice President for Alumni and Global Advancement, Nanyang Technological University, Singapore.

References

1. Interdisciplinary Graduate Programme. In: Nanyang Technological University. http://gc.ntu.edu.sg/Programmes/IGP/Pages/Interdisciplinary-Graduate-Programme-Scholarship.aspx. Accessed 20 Jan 2020
2. Graduate College. In: Nanyang Technological University. http://gc.ntu.edu.sg. Accessed 20 Jan 2020

3.4.4   Institutionalizing the New Curriculum

When connecting knowledge from various fields, the evolving curricula will inevitably bridge existing departmental or disciplinary boundaries. Courses and contributors will be in different departments, schools, or faculties. In the traditional university, the organizational boundaries tend to follow disciplinary structures, which are also reflected in academic career paths. Hence, a barrier to developing and institutionalizing the evolving curriculum is how it can be organized, and how it can achieve status within the academic structures.

There are some challenges that need to be handled wisely to afford the new curriculum sufficiently stable conditions. An inter-organizational curriculum may be necessary in the start-up phase. However, long-term sustainability may be easier to achieve within one organization. The strongest way to make new offerings more permanent is to appoint new faculty as responsible for the evolving curricula. Even one or two professors, especially senior ones, can often attract a critical mass to support a new curriculum.

If there are mechanisms to allow the creation of developmental ad hoc offerings, some of these can then evolve into recognized new subjects, finally approved as part of a curriculum. A key consideration is how the offering will be resourced. An important indicator of sustainable curriculum is when it is funded as a regular offering by the university. There is much potential return on investment, both from a scientific point of view, and considering the potential for knowledge exchange and innovation. Therefore, it would be appropriate for university leaders to invest in these various forms of evolving curriculum.

3.5   Preparing for Innovation

PREPARING FOR INNOVATION

3.5.1   Making Young Professionals Successful

Taken together, the previous three practices form a basis for a contemporary education. *Integrated Curriculum* addresses the integration of disciplinary fundamentals
with the essential skills for working life. *Teaching for Learning* provides further
considerations for developing deeper working understanding, self-efficacy, and
self-learning abilities. *Education in Emerging Thought* exposes students to cutting-
edge and cross-disciplinary thought. But there still remains a gap between this
extensive preparation and a readiness to be an effective young innovator. This gap is
addressed by a practice called *Preparing for Innovation*.

Box 3.8 Goals of Preparing for Innovation

Universities will educate students with a deep working knowledge of
fundamentals while better preparing them as agents of knowledge exchange
and innovation
**By implementing educational activities for leadership, management
and entrepreneurship,**
*educating students who are better prepared for their potential roles
in innovation.*

Innovation is a complex and multifaceted process [20, 21]. Successful
technology-based innovators need to be deeply knowledgeable about the technology and effective as agents of knowledge exchange. They need to understand the
nature of the market or social need that is addressed. They need to understand the
life cycle of an innovative good or service. They need to understand entrepreneurial
venture creation. They need to know how to manage technology in a sector. They
need to effectively establish a knowledge organization. They need to be able to lead,
often in an environment of uncertainty and change.

Traditionally, it was thought that students would develop this expertise on the job.
But young entrepreneurs and innovators are more likely to be successful in their early
ventures when they have some formal exposure to the relevant process knowledge in
their education [22]. To the extent that we propose that students should learn about innovation, the specifics have to be molded to the educational context. Some of these topics
can be introduced at the undergraduate level, but most of this learning is more appropriately addressed at the postgraduate level, and increasingly by learners with work experience—much as MBA students, many with work experience, learn in business school.

To effectively prepare students for innovation, we do not propose an approach
based on traditional readings and lectures. Instead, we recommend that students
start the learning sequence by exploring a motivating need, and then getting exposed
to models and a modest amount of theory. This is directly followed up with active
learning—case studies, role playing, or project work. The real learning occurs when
the students *reflect* on these experiences, on their processes and methods, as well as
their successes and setbacks. Reflection is guided by a mentor or advisor, and often

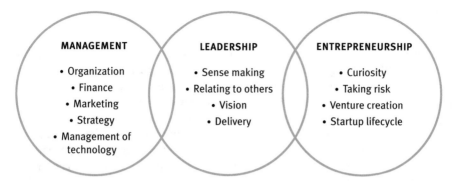

Fig. 3.12 Elements of the practice of preparing for innovation

deepened by peer feedback. Needless to say, the educators may need preparation and support to effectively deploy this model of teaching and learning.

Preparing for Innovation supports learning across a spectrum, including elements that are conventionally called innovation, entrepreneurship, leadership, and management of technology. There are important attitudes to be honed: curiosity, integrity, resiliency, commitment, and the ability to assess and take appropriate risk (Fig. 3.12).

Three key actions recommended to prepare students for innovation are shown in Box 3.9.

Box 3.9 Preparing for Innovation—Practice and Key Actions

Preparing for Innovation supports student development of the advanced skills and knowledge needed to become effective innovators: leadership, management, innovation, and entrepreneurial skills, and an aptitude to be curious, identify opportunities, and take appropriate risk. These skills are acquired through: exposure to models and theory; cases, scenarios, and projects; and mentoring and reflection.

Rationale: Innovation is a complex and multifaceted process, requiring individuals and teams with many skills beyond technical knowledge, and the confidence to apply them. Preparing young innovators and entrepreneurs, this way will increase the likelihood of early successful ventures with efficient time-to-market.

Key actions:
- Learning in innovation and entrepreneurship which focuses on how to support the entire product life cycle, along with the creation of a new ventures.
- Learning how to manage the development and deployment of technology, within a specific technology and market sector.
- Learning the skills to make sense of complex situations, rally others, create visions, and work relentlessly to deliver solutions that address common goals.

The key outcome is talented graduates, ready to participate in innovation.

3.5.2   Education in Innovation and Entrepreneurship

The innovator faces many challenges related to the product life cycle: identifying market needs; deriving product strategy; conceptual and detailed design; implementation, integration, and validation; marketing and sales; deployment, service, and retirement. An innovator who is also an entrepreneur is simultaneously creating a product and an enterprise, which requires additional expertise [23, 24]. This includes assembling the initial team, establishing the company, developing the business plan and capitalizing the company, and managing intellectual property. All technical, business, and marketing processes have to be started from zero. This demands preparation considerably beyond the essential life and professional skills discussed in the practice on curriculum.

Our model of acquiring the attitudes required for innovation and entrepreneurship is based on extending the concept of *self-efficacy* [15], introduced above in Teaching for Learning. We extend it to include the specific self-confidence that makes a young innovator believe themselves capable of succeeding in product and venture creation. Students need to develop key attitudes that include curiosity, propensity for taking action, and confidence to assess and take appropriate business risk.

The preparation for innovation and entrepreneurship that helps student develop the experience, skills, and attitudes would ideally include:

(a) Increasingly authentic projects that prepare students to implement new products, services, and systems and to launch new ventures.
(b) Repeated practice at identifying and addressing ambiguous and unstructured problems, developing solutions and advocating their adoption.
(c) Experiences in other essential aspects of innovation and entrepreneurship, such as manufacturing, services, and enterprise development.
(d) Ongoing discussions with real innovators and entrepreneurs, who share reflections on their experiences and provide mentoring to the students.

As an example, at Stanford University, the Stanford Technology Ventures Program (STVP) offers a wide range of opportunities for innovation and entrepreneurship experiences, attracting students from different disciplines, and producing successful entrepreneurs (Case 3.10).

Case 3.10 Stanford University—Stanford Technology Ventures Program

A center that excels at entrepreneurial education, with a specific emphasis on skill-building and scalable technology ventures, can increase its impact if its activities are rooted in scholarship on entrepreneurship and innovation, and contextualized for local ecosystems.

Established in 1996, the Stanford Technology Ventures Program (STVP) is the entrepreneurship center within the School of Engineering at Stanford University [1]. Part of the Department of Management Science and Engineering, STVP engages students across all disciplines to accelerate principled entrepreneurship

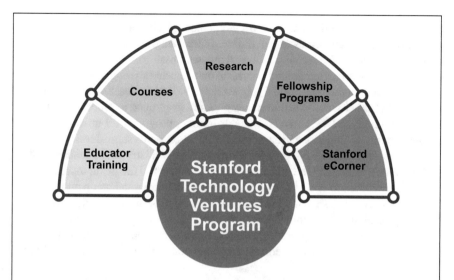

Fig. 3.13 Major efforts of the Stanford Technology Ventures Program

education and research, and support a new generation of leaders to address twenty-first century opportunities. Even in the highly decentralized campus entrepreneurial ecosystem, 10–15 percent of Stanford students annually engage with STVP courses and programming.

STVP's portfolio of 35 affiliated courses covers a breadth of topics to develop entrepreneurial mindsets and complement students' discipline-specific educations. Many courses also prepare students to leverage emerging technologies to create organizations and successfully scale them over time (see Fig. 3.13). Insights from STVP's robust research group (with 40 PhD alumni) directly influences the insights shared in the courses, which are taught by a dynamic community of tenure-line faculty and practitioner lecturers. STVP has also been the home of the first lean start-up methods course taught by Steve Blank, and derivative courses on solving national security problems (Hacking for Defense), and required courses for NSF grantees interested in commercialization (I-Corps).

For students, some of STVP's key offerings are immersive cohort programs. There are 400 combined alumni of these programs, with the majority having worked at, or founded, technology ventures, and including some who led prominent and successful ventures. The original example of these offerings is the Mayfield Fellows Program, an intensive, 9-month entrepreneurial leadership program for 12 undergraduate students from diverse majors. Participating students develop skills and experience through courses, mentoring, a Silicon Valley venture internship and the opportunity to build cases on these combined experiences. STVP also offers additional cohort experiences for engineering master's and PhD students, offering a front-row seat to innovation and venture creation through workshops, soft-skill development, and industry field experiences to help students recognize opportunities and tackle entrepreneurial challenges. STVP has also piloted programming for faculty at different career stages to gain understanding of how to commercialize technology-based innovations.

STVP maintains a committed interest to outreach and sharing resources beyond Stanford. With more than 3000 freely available videos, podcasts, and articles, the Stanford eCorner website [2] offers insights and inspiration to global educators.

Additionally, STVP delivers educator training programs for universities and governments to support entrepreneurial education ecosystem development. A recent example is STVP's collaboration with the Ministry of Education in the United Arab Emirates. The project trained more than 100 faculty at more than 40 UAE universities to scale educational capacity for entrepreneurship and innovation, while creating shared language and skills for university students across the country. Beyond the significant impact on a global community of educators, STVP's outreach work provides its own scholars, educators, and students with vital global context for how entrepreneurship continues to evolve in a technology-driven world.

Prepared based on input provided by Tom Byers, Entrepreneurship Professor in the School of Engineering, Stanford University; and Matthew Harvey, Executive Director, Stanford Technology Ventures Program, Stanford, California, USA.

References

1. Stanford Technology Ventures Program. https://stvp.stanford.edu. Accessed 20 Jan 2020
2. Stanford eCorner. https://ecorner.stanford.edu. Accessed 20 Jan 2020

3.5.3   Education in the Management of Innovation

In addition to the preparation for innovation, students should learn the skills of management—the control of an organization to reliably meet objectives within available resources. At the core of a program in management of innovation are the foundational topics of management like marketing, accounting, finance, and organization behavior. As a program in the Management of Innovation is specialized for students who will be involved in managing ventures based on new technology and emerging thought, this might suggest adding topics like entrepreneurial finance and product marketing.

Some Management of Technology programs link to either an *aspect of product development* or an *emerging technology*. The variant of the program emphasizing product development will combine business knowledge with preparation in product or system development. This linkage can be to the upstream aspects of product or system design or to the downstream aspects of manufacturing, supply chains, and operations. These programs typically admit postgraduate students with several years of relevant job experience.

In the other variant emphasizing emerging technology, students learn about a science or technology at level beyond an undergraduate preparation. They learn about the technology from the perspective of how it could create value in new products. They also learn about the relevant industrial and market sectors and their potential to accept innovation. These programs are often at the master's level.

The curriculum of such programs might include a taught component and a project component, on-campus or as an internship. The curriculum would contain:

(a) Technical education based on an emerging field of thought or technology
(b) Advanced product development skills, including system design and operations
(c) Core management disciplinary materials

As an example, the University of Cambridge offers graduate programs to prepare students for commercializing knowledge at the frontiers of research, developing the skills needed for effectively managing the innovation process and starting new ventures (Case 3.11).

Case 3.11 The University of Cambridge—Masters in Management and Technology

Blended graduate programs that include elements of cutting-edge science and technology, along with business education, prepare young managers of the innovation process.

The students of today are the innovators, entrepreneurs, and managers of tomorrow. The purpose of the graduate programs developed by the Cambridge-MIT Institute (CMI) was to enhance the knowledge, skills, and innovative capability of students. In particular, they were intended to prepare students to use their knowledge of what is happening at the frontiers of research, to commercialize it, and turn it into new enterprises.

The CMI programs were based on a model for a 1-year taught MPhil (Masters of Philosophy) that combined advanced technical material with skills and knowledge of management and innovation. They filled a gap between technically specialized master's programs and general management programs such as an MBA. The design of the programs was influenced by the MIT professional practice programs in Leaders for Global Operations (which worked between Management and Engineering) and Biomedical Enterprise (which worked between Management, and Health Sciences and Technology).

A total of six programs were created, in technical areas ranging from chemical engineering to nanotechnology and bioscience. The technical content assumed a strong undergraduate degree preparation in the topic. The programs combined exposure to the latest cutting-edge science, with an emphasis on pathways to application and commercialization.

The management content introduced students to the skills required for managing innovation and running businesses. Areas covered included traditional

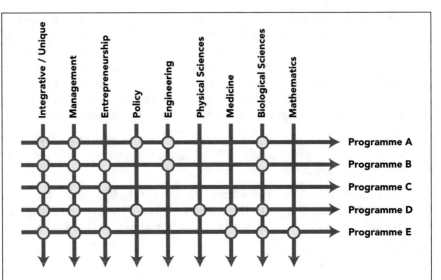

Fig. 3.14 Open architecture and platform sharing for programs

management topics; for example, marketing, finance and organizational behavior, but all with an emphasis on how these areas could support innovation. Elements of entrepreneurship (e.g., starting a company, capitalization, licensing) and policy, including ethical, legal, and regulatory issues, were also covered.

The suite of programs was based on the principle of open architecture. In particular, the management component was constructed as a platform of modules offered to all of the programs simultaneously (see Fig. 3.14).

The science and management content was integrated through workshops and case studies and through a major in-company, team-based consulting project in a venture relevant to the particular technical specialty.

As an example, consider the MPhil in Bioscience Enterprise [1]. As described on its website, it is a "multi-disciplinary biotechnology and business degree course at the University of Cambridge, designed for high-achieving individuals with an enthusiasm for enterprise. Students extend their knowledge of the latest advances in exploitable biotechnology and medical science, and gain an understanding of the ethical, legal and regulatory issues associated with bringing scientific advances to market. The course equips students with the requisite skills in enterprise, management and entrepreneurship required to take up executive roles, establish and build biotechnology companies or work in the life sciences consultancy sector."

A second example is the MPhil in Micro and Nanotechnology Enterprise [2]. "The program is intended for those with a good first degree in the physical sciences and engineering who wish to develop research skills and a commercial awareness in micro- and nanotechnology. It combines cutting-edge science with business practice skills, giving students knowledge and experience of a range of disciplines. This should enable students graduating from the course to evaluate the scientific importance and technological potential of new developments in the field of micro- and nanotechnology and provides an unparalleled educational experience for entrepreneurs in these fields."

The programs have been running since 2004, and each has educated hundreds of students.

Contributed by Jochen Runde, Professor, the Judge Institute of Business, the University of Cambridge, Cambridge; and Nick Oliver, Professor, University of Edinburgh Business School, Edinburgh, UK.

References

1. MPhil in Bioscience Enterprise. In: Department of Chemical Engineering and Biotechnology. University of Cambridge. https://www.ceb.cam.ac.uk/postgraduates-tab/mphil-mbe. Accessed 20 Jan 2020
2. MPhil in Micro and Nanotechnology Enterprise. In: Department of Materials Science & Metallurgy. University of Cambridge. https://www.graduate.study.cam.ac.uk/courses/directory/pcmmmpmne. Accessed 20 Jan 2020

3.5.4   Scientific and Technology Leadership Education

Leadership is distinct from both innovation and management [25, 26]. It is commonly defined as the processes whereby an individual influence a group to accomplish shared objectives, often associated with change or transformation. Leadership is not always tied to a position, but can be informal, personal, and distributed [27, 28]. It occurs in various domains: education, business, politics, culture. Our focus here is leadership in scientific and technical innovation, preparing students to be able to mobilize internal and external resources and lead teams.

In our view, leadership skills can be developed. Education for leadership aims to develop students' *knowledge about* leadership, develop their *leadership skills* through personal experiences, develop their *character* and *identity*, and their *understanding of the context* [29]. This can be done by giving students practice in making sense of complex situations, rapidly integrating knowledge and probing actively to understand issues more completely. Students need to learn to relate to and rally others [30], advocating the shared objectives. They need to create visions of the desirable outcomes of an effort, and then work intensely to deliver on those visions [31].

To teach *innovation leadership*, the suggested approach is to mix scenario-based practice, leadership theory, and guided reflection. This is done in a realistic context of activities that supports students in cultivating their capabilities, includes participating in and leading teams to deliver new products.

Education for leadership can include:

(a) Exposure to theory, analytical concepts, and frameworks for understanding the capabilities and skills of leadership.

(b) Group-based and highly immersive leadership activities designed to challenge student assumptions and develop their leadership skills.
(c) Various forms of feedback guiding students through reflection and self-assessment on their performance and identifying how to improve.
(d) Focus on bringing about change, transformation, and innovation.

Some universities have specialized leadership programs. At the University of Toronto, engineering students are offered a learning environment to prepare them for *leading change to build a better world*, through a sequence of learning activities connected across the curriculum (Case 3.12).

Case 3.12 University of Toronto—Troost Institute for Leadership Education in Engineering

Life changing opportunities for leadership exploration can be provided across embedded modules, dedicated courses, and co-curricular events.

Innovation is not just about technology, it is about change. The capacity to lead change distinguishes those twenty-first century technical professionals who will generate significant transformations. The University of Toronto Troost Institute for Leadership Education in Engineering (Troost ILead) was founded in 2010 with the vision: *Engineers leading change to build a better world.*

The Institute provides students with leadership knowledge and development opportunities. It provides a learning environment where personal growth is empowered, creativity is encouraged, humility guides learning, diverse perspectives are invited, and exuberance is contagious. Students explore various leadership frameworks and meditate on their own leadership styles through a series of workshops, team projects, and reflection activities. Leadership skills are built on these strengths to make students more effective engineers [1].

The Institute develops integrated curricular approaches: co-curricular and specialty curricular programming for leadership education for both undergraduate and graduate students in engineering (see Fig. 3.15). All aspects of the program are based on a leadership development model that starts with self-awareness and develops through cycles of action and reflection at the team, organizational and societal level (see Fig. 3.16). Affiliated faculty conduct research on the pedagogy of leadership education, and on leadership practice in engineering-intensive enterprises, producing tools and frameworks used across the program.

Leadership content is integrated into core curricular design courses, with a focus on building effective teams. All first-year students receive at least 6 h of instruction on self and team performance, and use the Team Effectiveness Learning System tool for structured team feedback. Although this leadership educational model is personalized, ILead has been able to effectively scale the program benefitting more than 2000 students this year.

Positive experiences in the first-year curriculum draw students into co-curricular and specialty offerings. Students learn how to effectively handle complex,

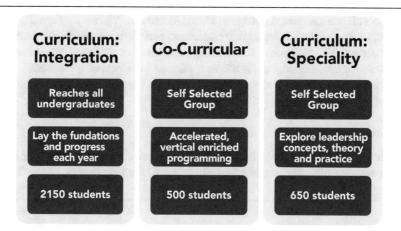

Fig. 3.15 Curricular and co-curricular foundations to build a comprehensive program (numbers of students involved are annual)

Fig. 3.16 Troost ILEAD Leadership Development Model

human challenges that often mean the difference between success and failure. In 2019, more than 900 students participated in co-curricular activities for undergrad and graduates, including drop-in Leadership Labs, summer fellowships, and roundtables, where small group coaching is given to club leaders. Our OPTIONS program offers professional and career exploration for cohorts of PhD students who want to explore nonacademic careers.

A further 650 students were enrolled this year in specialty courses. The curriculum covers topics in six undergraduate and eight graduate elective courses, such as Foundation of Engineering Leadership, Cognitive and Psychological Foundations of Effective Leadership, and The Power of Story: Discovering your Leadership Narrative. A strength of ILead is its research-based treatment of ethics, which shapes the subject called The Art of Ethical & Equitable Decision-Making in Engineering that and helps students navigate successfully through ethical dilemmas in engineering.

> Troost ILead has created a program around a virtuous cycle: deeply engaging with industry to understand its leadership challenges; performing social science research on issues and teaching approaches; creating educational modules and courses to support learning of leadership; and graduating students to meet society's needs.
>
> Contributed by Emily Moore, Director, Troost ILead, University of Toronto, Toronto, Canada.
>
> **Reference**
>
> 1. Reeve D, Evans G, Simpson A et al (2015) Curricular and co-curricular leadership learning for engineering students. Collected Essays Learn Teach 8:41–56

3.6   Chapter Summary

In this chapter, we argue that when a university wants to strengthen its contributions to sustainable economic development, then *education really matters*. We present four systematic practices for educating students as effective knowledge exchange agents (Fig. 3.17). These practices make tangible the principle of systematic knowledge exchange (Box 2.3).

Integrated Curriculum is the practice of designing curricula based on a vision of what the graduates should be able to do, informed by the needs of partners. The educational aims include a deep working understanding of fundamentals together with the essential skills, approaches, and judgment needed in working life. We strive to achieve progression and connections throughout the curriculum, also complemented with co-curricular activities. This is the foundational practice and supports the other educational practices.

Teaching for Learning concerns designing courses and projects to achieve the desired learning outcomes formulated in the Integrated Curriculum practice. Learning and assessment activities should be aligned with the intended learning outcomes, forming a motivational context. This involves active and experiential learning methods including various forms of digital resources. The educational experience should also further students' self-efficacy and capacity for independent learning.

Education in Emerging Thought is a practice connecting education and research by promptly transferring new knowledge to the curriculum in various ways, from creating new programs to flexible course structures. The focus is on discoveries at the frontiers or intersections of disciplines and knowledge linked to emerging needs in society and enterprises. The primary outcome is graduates with cutting-edge knowledge, prepared to work in cross-disciplinary environments.

Preparing for Innovation is the practice of educating students for innovation and entrepreneurship, and for management and leadership in scientific and technical innovation endeavors. Drawing on the Integrated Curriculum and Teaching for Learning practices, we emphasize the need to place students in authentic situations,

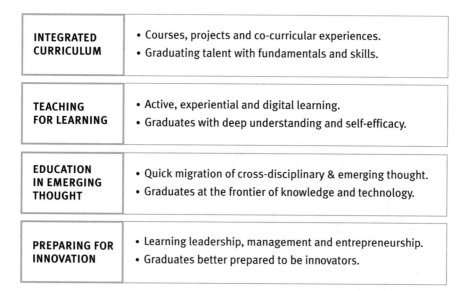

INTEGRATED CURRICULUM	• Courses, projects and co-curricular experiences. • Graduating talent with fundamentals and skills.
TEACHING FOR LEARNING	• Active, experiential and digital learning. • Graduates with deep understanding and self-efficacy.
EDUCATION IN EMERGING THOUGHT	• Quick migration of cross-disciplinary & emerging thought. • Graduates at the frontier of knowledge and technology.
PREPARING FOR INNOVATION	• Learning leadership, management and entrepreneurship. • Graduates better prepared to be innovators.

Fig. 3.17 Educational practices, processes, and outcomes

and to further learning through reflection on their experiences and feedback. The primary outcome is graduates ready for innovation.

These practices are relevant for furthering the impact of a university because the graduates constitute the major form of knowledge exchange with partners in industry, enterprise, and government organizations. From the perspective of the students, they should expect no less from their education than to become equipped for handling challenges and accessing new opportunities. This will benefit themselves, their partners and employers, and a sustainable society.

The themes of sustainable development are woven throughout these practices. The learning outcomes can include approaches to sustainability, which can also be a strong motivational context for learning. Much of sustainable development is at the intersections of existing disciplines. Sustainability requires innovation, management, and leadership.

References

1. Seely B (2005) Patterns in the history of engineering education reform: a brief essay. In: National Academy of Engineering, Engineer of 2020: National Education Summit. National Academies Press, pp 114–130
2. Crawley E, Malmqvist J, Östlund S, Brodeur DR, Edström K (2014) Rethinking engineering education: the CDIO approach, 2nd edn, Springer Cham
3. MIT (1949) Committee on educational survey. Report to the Faculty of the Massachusetts Institute of Technology (The Lewis Report). Technology Press, Cambridge, MA

4. Barrie SC (2006) Understanding what we mean by the generic attributes of graduates. High Educ 51:215–241
5. Edström K (2018) Academic and professional values in engineering education: engaging with history to explore a persistent tension. Eng Stud 10:38–65
6. Wankel LA, Wankel C (2016) Integrating curricular and co-curricular endeavors to enhance student outcomes. Emerald Group Pub Ltd, Bingley
7. Feisel L (1986) Teaching students to continue their education. In: Proceedings of the Frontiers in Education. Arlington, Texas
8. Biggs J, Tang C (2011) Teaching for quality learning at university. Open University Press
9. Marton F, Säljö R (2005) Approaches to learning. In: Marton F, Hounsell D, Entwistle NJ (eds) The experience of learning. Scottish Academic Press, Edinburgh, pp 36–55
10. Boud D, Falchikov N (2007) Rethinking assessment in higher education: learning for the longer term. Routledge
11. Kolb DA (2014) Experiential learning: experience as the source of learning and development, 2nd edn, Pearson Education, Upper Saddle River, NJ
12. Freeman S, Eddy SL, McDonough M, Smith MK, Okoroafor N, Jordt H, Wenderoth MP (2014) Active learning increases student performance in science, engineering, and mathematics. Proc Natl Acad Sci 111:8410–8415
13. Mazur E (1996) Peer instruction: a User's manual. Pearson Education
14. Roediger HL III, Karpicke JD (2006) The power of testing memory: basic research and implications for educational practice. Perspect Psychol Sci 1:181–210
15. Bandura A (1997) Self-efficacy: the exercise of control. Worth Publishers
16. Siemens G, Conole G (2011) Connectivism: design and delivery of social networked learning. Special Issue International Review of Research in Open and Distance Learning. http://www.irrodl.org/index.php/irrodl/article/view/994/1820. Accessed 2 Jan 2020
17. Ambrose SA, Bridges MW, DiPietro M, Lovett MC, Norman MK, Mayer RE (2010) How learning works: seven research-based principles for smart teaching. Jossey-Bass, San Francisco, CA
18. Boud D, Molloy E (2013) Feedback in higher and professional education: understanding it and doing it well. Routledge, Abingdon
19. Gibbons M, Limoges C, Nowotny H, Schwartzman S, Scott P, Trow M (1994) The new production of knowledge: the dynamics of science and research in contemporary societies. Sage Publications, London
20. Roberts EB (1991) Entrepreneurs in high technology: lessons from MIT and beyond. Oxford University Press, New York, NY
21. Christensen CM, Roth EA, Anthony SD (2004) Seeing what's next: using the theories of innovation to predict industry change. Harvard Business School Press, Boston
22. Byers TH, Dorf RC, Nelson A (2015) Technology ventures: from idea to enterprise. McGraw-Hill, New York, NY
23. Dodgson M, Gann DN, Phillips N (2014) The Oxford handbook of innovation management. Oxford University Press, New York
24. Burlgeman RA, Christensen CM, Wheelwright SC (2009) Strategic management of technology and innovation. McGraw-Hill, New York
25. Nohria N, Khurana R (2010) Handbook of leadership theory and practice. Harvard Business School Press, Cambridge, MA
26. Hill LA, Brandeau G, Truelove E, Lineback K (2014) Collective genius: the art and practice of leading innovation. Harvard Business Review Press, Cambridge, MA
27. Ancona D (2005) Leadership in an age of uncertainty. MIT Center for eBusiness 6:1–4
28. Ancona D, Bresman H (2007) X-teams: how to build teams that Lead, innovate and succeed. Harvard Business Review Press, Cambridge, MA

29. Snook S, Nohria NN, Khurana R (2011) The handbook for teaching leadership: knowing, doing, and being. Sage Publications
30. McKee A, Boyatzis RE, Johnston F (2008) Becoming a resonant leader: develop your emotional intelligence, renew your relationships, sustain your effectiveness, kindle Edi. Harvard Business Review Press, Boston
31. Graham R, Crawley E, Mendelshon BR (2009) Engineering leadership education: a snapshot review of international good practice. White Paper sponsored by the Bernard M Gordon-MIT Engineering Leadership Program

Chapter 4
Research and Knowledge Exchange

4.1 Introduction

4.1.1 The Objectives of Research

Research is about discovery at the frontiers of knowledge, and the quest for increased understanding of our world. The importance of research to society is enormous. If effectively communicated, research discoveries have many kinds of impact. They reveal new knowledge about our society and universe and can inspire the citizenry. They inform the curriculum and help educate our students. Research contributes discoveries that can lead to innovative creations, new products and systems, and, eventually, to economic development.

> **Box 4.1 Objectives of Research and Knowledge Exchange**
>
> The general objective of research is to make *discoveries*—often revealing phenomena or truths that have previously existed but were unknown or unexplained. Discoveries can be new knowledge, facts, data, theories, models, analyses, and predictions and can be in single disciplines, as well as across disciplines (Fig. 2.1).
>
> Knowledge about discoveries is exchanged through formal and informal discussions, and publications. More proactive approaches include joint projects, personnel exchange, and involvement with professional development. And, graduates carry knowledge of discoveries to future work (Fig. 2.4).

We use the word *discoveries* to distinguish research outcomes from the synthesized outcomes of catalyzing innovation that we call *creations* (Chapter 5), and from the more general term, *knowledge*, which we apply to the outcomes of education and catalyzing innovation as well.

© Springer Nature Switzerland AG 2020
E. Crawley et al., *Universities as Engines of Economic Development*,
https://doi.org/10.1007/978-3-030-47549-9_4

Viewed through the lens of economic development, the specific objective of research is to *make discoveries at the frontiers of knowledge that have the potential for becoming more impactful instruments of knowledge exchange and innovation.* Good research has an *impact*: on other scholars, on those addressing the issues of society, or both. The potential for discoveries to become more impactful on innovation depends on the choice of the research topic, the way the research is conducted, what partners are involved, and how results are exchanged.

In Chapter 2, we discuss the important role research plays in economic development. Economic development occurs because of accelerating innovation and entrepreneurship. These follow as a result of enhancing knowledge exchange between the university and its partners: industry, small and medium enterprise, and government organizations. The two key principles of Chapter 2 are that:

- Knowledge exchange will become more effective when there is *a systematic approach* that carefully identifies needs of the partners, conducts university activities with a sensitivity to these needs, and proactively exchanges the university outcomes with partners.
- The knowledge that is exchanged emanates from the cross-disciplinary and integrated activities at a university—education *and* research *and* catalyzing innovation.

Now, our task is to examine the research component of these integrated and systematic activities, understanding that research is not isolated from education or catalyzing innovation [1].

4.1.2   Background and Opportunity

Universities are particularly suited to research at the *frontiers*—exploration at the limits of existing knowledge. Research at the frontier can include extending fundamental understanding. But research at the frontier can also be motivated by the need to work around fundamental limitations that become evident in the process of Maturing Discoveries. Topics at the frontier are also sometimes revealed when a team is developing directly implementable solutions to the needs of society, as is done in a Centre of Research, Education, and Innovation.

There are many pathways between research and final impact, necessitating a wide range of approaches. At a university, research emerges from the collective action of individual faculty. Once hired, faculty should have the freedom to choose their research topics and pathways. Their choice is influenced by personal interests, ambition, and institutional culture. The desire for peer recognition can be a progressive or conservative force. Researchers might also consider public priorities articulated by funding agencies, or by the needs of industry, enterprise, and government institutions.

The most salient features of good research are the selection of an important problem, and the dedication, passion, and curiosity on the part of good people. Discoveries should be driven by the pursuit of excellence and conducted in a meritocracy. Not surprisingly, these qualities are emphasized by funding agencies.

Table 4.1 The research practices

Practice icon	Practice name: description of the practice > and its outcome
IMPACTFUL RESEARCH	*Impactful fundamental research*: Pursuing fundamental new discoveries along a spectrum from curiosity-driven to use-inspired > yielding new knowledge with an impact on scholarship and society
COLLABORATIVE RESEARCH	*Collaborative research within and across disciplines*: Collaborating with other scholarly researchers within the university and externally > making discoveries that cross-disciplinary boundaries and are in new fields of thought
CENTRES	*Centres of research, education, and innovation*: Empowering larger scale integrated Centres of Research, Education, and Innovation > producing directly implementable and impactful solutions that address significant issues of society
STUDENT RESEARCHERS	*Undergraduate and postgraduate student researchers*: Engaging undergraduate and postgraduate students in research > energizing research, and yielding effective young researchers and agents of knowledge exchange

4.1.3   The Academic Practices of Research

A university can make discoveries that have the potential for becoming more effective instruments of knowledge exchange and innovation, by applying the four research practices presented in Table 4.1. The chapter that follows discusses these practices. Included are eight cases that help to frame and illustrate the practices.

The way these practices can change a university is demonstrated in Table 4.2, showing the transition from the reference situation—the average state of good universities—to the aspirational one.

Table 4.2 Reference and aspirational situation for research

Reference situation for research	Aspirational situation for research
Faculty characterize their research as fundamental and curiosity-driven, and are rewarded largely for their impact on other scholars	Faculty willingly carry on fundamental research on a spectrum from curiosity-driven to use-inspired, with potential impact on scholars and on those addressing the needs of society. [Impactful Research]
Faculty prize their independence and tend to work alone or in small informal groups of disciplinary peers	Faculty prize their independence but are more likely to work collaboratively with colleagues to make cross-disciplinary and broader contributions [Collaborative Research]
Faculty do not normally have the opportunity to work across university boundaries on large-scale efforts	Faculty spend some of their time working in larger scale efforts, joining distinguished industry practitioners, and others from outside the university in developing solutions to societal issues [Centres]
Faculty carry on research work primarily with doctoral and postdoctoral scholars	Faculty work with and inspire contributions from teams whose members range from young undergraduates to postgraduates and postdoctoral scholars [Student Researchers]

The four practices presented in Table 4.1 will help a university in its evolution towards these four aspirational situations. The main distinguishing features of these practices are that: the first is about fundamental *impact*; the second is about increasing *scope* through collaboration; the third about *directly implementable outcomes* to problems of society. The final one is about engaging and enhancing students' *capacity and capability*. Other practices relevant to research are Education in Emerging Thought, which migrates new and cross-disciplinary discoveries to the education (Chapter 3), and Maturing Discoveries and Creations, which transforms discoveries into innovative creations (Chapter 5).

4.2   Impactful Fundamental Research

4.2.1   A Spectrum of Impactful Fundamental Research

This first research practice provides guidance on the conduct of fundamental research, which can lie along a spectrum from curiosity-driven to use-inspired. Because effective research can have an impact on other scholars and on those who address societal or innovation issues, we consider it the *foundational practice* of Research and Knowledge Exchange.

Box 4.2 Goals of Impactful Fundamental Research.

Universities and their researchers will make discoveries that are more effective instruments of knowledge exchange and innovation,
By pursuing fundamental new discoveries along a spectrum from curiosity-driven to use-inspired,
Yielding new knowledge with an impact on scholarship and society.

The traditional model represents research on a one-dimensional axis, with basic research on one end and applied research towards the other (Fig. 4.1). With time, basic research is expected to lead to applied research and then development, implying all new products and system spring from scientific discovery. By comparison with practice, this model is oversimplified [2].

In the latter half of the twentieth century, an alternative view was emerging. Harvey Brooks of Harvard wrote "… the terms basic and applied are, in another sense, not opposites. Work directed towards applied goals can be highly fundamental in character in that it has an important impact on the conceptual structure or

Fig. 4.1 Traditional one-dimensional *(upper)* and two-dimensional models of research, indicating that fundamental research can be curiosity-driven or use-inspired *(adapted from Stokes 1997)*

outlook of a field. Moreover, the fact that research is of such a nature that it can be applied does not mean that it is not also basic [3]."

Stokes proposes a solution to this apparent contradiction [2]. He suggests viewing research on a two-dimensional grid, with the vertical axis representing the quest for *fundamental understanding*, and the horizontal axis showing the degree of *consideration of use* (Fig. 4.1). In this practice, we concentrate on research aimed at more fundamental understanding, and organize this effective practice around the second axis as shown by the shaded quadrants. In this way, the spectrum of more *fundamental research* ranges from curiosity-driven to use-inspired. Stokes called these endpoints Bohr and Pasteur [2].

In curiosity-driven fundamental research, scholars are motivated by interesting problems, which may or may not be immediately relevant to currently perceived needs. They can benefit from collaboration with like-minded scholars, but otherwise engagement may be low. Conant called this "uncommitted exploration of a wide area of man's ignorance [4]."

This kind of research advances knowledge and can lead to unexpected discoveries with far-reaching impact. It is the seed corn for the future. In time, these unexpected discoveries can inform use-inspired research and migrate through translational research. In the mid- to long term, they can impact industry, enterprise, and government institutions through knowledge exchange, and therefore influence economic development. Stokes named this Bohr's quadrant. More recent examples would be:

- The American Townes, and Russians Bazov and Prokhorov, who received the 1964 Nobel Prize for the laser, which they developed as an experiment in the amplification of light, but resulted in significant economic impact on communications and manufacturing.
- The Australian John O'Sullivan who was working on an algorithm for reducing data from radio astronomy experiments, but in 1994 received a patent for what went on to be an important algorithm in WIFI, an indispensable element of our economy.

Use-inspired fundamental research also seeks discoveries at the frontier, but is motivated by a problem of industry, enterprise, government, or society. In use-inspired research, scholars learn about the needs of society and industry by scanning for addressable issues of a fundamental nature. Engagement with industry and society may not be deep, but is sufficient to reveal the challenging research questions. Conant named this "a research program aimed at a specific goal [5]."

Use-inspired research can also lead to unexpected fundamental discoveries and peer recognition. But since it was motivated by a need of society, its successful uptake by industry is more likely in the near term. It can impact economic development in the near- to mid-term. Stokes called this Pasteur's quadrant. More recent examples are:

- The Chinese Tu Youyou who did fundamental scholarly research that led to her 2015 Nobel Prize in medicine, but was always focused on solutions for malaria.

- The American Grace Hopper who, motivated by the need to program computers, developed higher level languages, and received the US National Medal of Technology in 1991.

In fact, these two approaches, curiosity-driven and use-inspired, anchor two ends of the spectrum of fundamental research. Faculty have the right and responsibility to select important problems along this spectrum and produce impactful results. An individual faculty member may have their research concentrated at one point in the spectrum, or spread in a portfolio along the spectrum. A spread portfolio has its benefits. Understanding of use can flow towards the curiosity-driven work, and new curiosity-driven discoveries can flow towards use-inspired work. A research group, or an entire university, can aspire to a balanced portfolio of projects on this spectrum.

At a national or supra-national level, investment can also range from curiosity-driven to use-inspired. This is an issue of finding the right balance, as illustrated in the two related cases. In Case 4.1 on the European Research Council, the EU assessed that its economic development would be strengthened by creating a funding stream that supports investigator-driven research based on scientific merit. They strengthened the curiosity-driven part of their spectrum. In contrast, the Australian government has put in place funding and incentives to attract larger teams to use-inspired projects. These address issues of society and innovation, as discussed in Case 4.2 on the Australian National University.

Case 4.1   The European Research Council (ERC)

A funding mechanism that supports investigator-driven research plays an important role in advancing science and supporting societal development.

The ERC's mission is to encourage the highest quality research in Europe through competitive funding and to support investigator-driven frontier research across all fields, on the basis of scientific excellence.

Frontier research is at the forefront of creating new knowledge and developing new understanding. The European Research Council's scope spans fields in the social sciences and humanities, the life sciences, and the physical and engineering sciences. The theme of the research is investigator driven, with the researchers identifying opportunities, without influence by governments or private interests. This makes the system more agile and responsive to breaking opportunities. The ERC awards are based on *scientific excellence*, "interesting, relevant and new," but are not required to reference how the science will be applied.

There is open and true competition for funding, and, unlike other EU schemes, each nation does not necessarily receive a share of funding in proportion to its contribution. In fact, one criticism is that some countries get smaller fraction than their contribution.

Today, the ERC issues multiyear grants to individual investigators. It is open to all levels of investigator. The applicant pool is segmented by experience level: Starting grants (2–7 years since PhD); Consolidator grants (7–12 years since PhD); and Advanced grants for those with significant research achievement in the last 10 years.

The ERC has become the gold standard for European research accomplishments. Universities now advertise their ERC count along with their Nobel and other major prizes. A principle benefit of the scheme is the recognition of *frontier research*. These results could lead to new and unexpected discoveries, which creates the foundation for the future of science, as well as basis of new industries, markets, and broader social innovations. The ERC provides an important recognition for individual scientists. It is especially important for younger scientists, flagging them up as high potential investigators, and serving as strong support at their next review.

The ERC strengthens European research, providing competition on an international scale. It provides benchmarking and allows nations to calibrate their effort and identify needs for funding. It raises the status and visibility of European research and attracts talent from abroad. It allows the EU to compete with other strong national systems and rationalize its science investments.

The origin of the ERC was in the early 2000s, when the economy was in recession and the jobless rate was high. European political leaders were aware of the strong investments in fundamental science by organizations like the US National Science Foundation and National Institutes of Health. They recognized that there was underfunding of the kind of fundamental research that leads to new discoveries and to new science-based industry. They created the ERC to invest in research that addresses emerging issues confronting society.

The investigator-driven ERC is designed to complement other EU funding programs that may involve more consideration of use, such as the framework programs and the Knowledge Integration and Innovation Centers, as well as programs of member states.

Since 2007, the ERC has funded 9000 projects and represents 17% of the overall Horizon 2020 budget (13 of 77 billion EUR) [1].

Prepared based on input provided by Ernst Ludwig Winnacker, founding Secretary General, European Research Council.

Reference

1. European Research Council. https://erc.europa.eu. Accessed 20 Jan 2020

Case 4.2 Australian National University (ANU)—A Shift towards Impact through Research and Innovation

ANU strives to continue its excellence in fundamental research, while embracing the government's initiatives encouraging universities to contribute to innovation.

Australia has strong economic fundamentals, but underperforms in technology-based innovation [1]. Australia's rate of collaboration between industry and researchers was reported as the lowest in the OECD's 36-country membership. On the other hand, Australia has a strong university system, with a tradition of research excellence. Australia has seven universities in the top 100 in the 2020 QS World University ratings.

The government sees an opportunity to better connect the assets of its universities with the needs of the Innovation and Science Agenda. It has instituted funding mechanisms and incentives to influence the actions of universities.

Australian universities receive a substantial fraction of their research funding from block grants, which includes an incentive to attract other research funding (see Fig. 4.2). For every 100 dollar of other government funding received, the block grant is topped-up by 30 dollars. This same formula has now been extended to industry funding.

The Australian Research Council (ARC), provides competitive funding through the Discovery Projects, targeted at excellent fundamental research. But, there is increasing emphasis on Linkage Projects that develop long-term strategic research alliances aimed at securing commercial and other benefits of research. Similar emphasis on clinical and applied impact is reflected in funding of the National Health and Medical Research Council, and the Commonwealth Scientific and Industrial Research Organization, the government lab system.

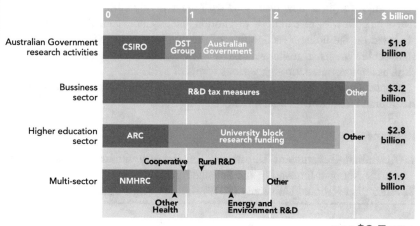

Fig. 4.2 Australian Government Investment in R&D, 2015–16

To assess the quality of universities, the ARC administers a broad-based evaluation called Excellence in Research for Australia (ERA). The 2018 ERA was coordinated with the first national Engagement and Impact assessment, which examined how universities translate their research into economic, environmental, and social benefits.

Clearly, the government is using funding and the assessment schemes to shift the universities to an external focus. How is it changing the culture and operation of the ANU, one of the country's leading research universities? ANU was founded as a research only institution in 1946, and in 1960 admitted undergraduate students. It is well positioned to undertake another transformation.

The ANU Strategic Plan [2] states that "In research, we must lead in breaking down the barriers between universities, society and industry. Equally our expectation must be that our research creates innovative outcomes contributing to the economic and general public good." ANU is investing in its Grand Challenges Scheme that funds interdisciplinary research projects that deliver solutions to national priorities, while strengthening core disciplines. One Grand Challenge is "Our Health in Our Hands," a 5-year, AUD 10 m interdisciplinary initiative on personalized health care from genomic, diagnostic, therapeutic, policy and public health perspective.

In addition, ANU has introduced Innovation Institutes to engage in translational collaborative research. The first is the 3Ai, drawing on an interdisciplinary team focused on Autonomy, Agency, and Assurance required for the adoption of AI into society.

To help the faculty develop broader relationships with industry, ANU has expanded its business development capability. Focusing on Strategic Partnerships and Projects that leverage ANU's expertise and networks is essential in building broader government and industry initiatives.

To align for these new directions with the evolving culture, ANU needed to revise its policy for researchers' promotion and advancement. The new policy is "focused on the holistic recognition of the quality, productivity and impact of staff achievements in research/creative activity, education, service and leadership, as demonstrated through various forms of evidence."

Contributed by Michael Cardew-Hall, Pro Vice Chancellor (Innovation), Australian National University, Canberra, Australia.

References

1. Australian Government Department of Industry Innovation and Science (2015) National Innovation and Science Agenda Report
2. Australian National University (2019) ANU Strategic Plan 2019–2022

Four key actions for implementing this practice are summarized in Box 4.3, and they are each discussed further below.

Box 4.3 Impactful Fundamental Research—practice and key actions.

Impactful Fundamental Research is a research process that seeks fundamental discoveries, whose motivation lies along a spectrum, from curiosity-driven to use-inspired. The main outcomes are discoveries that are broadly *impactful* on other scholars, on the issues of society and economy, or on both. This is the foundational practice for research.

Rationale: There are many pathways from research results to economic impact. It might seem that use-inspired research would have more direct near-term impact, but curiosity-driven research can also produce high-impact unexpected discoveries. A balanced portfolio of these approaches might more robustly address the needs of partners.

Key actions:
- Granting researchers the freedom and associated responsibility to undertake fundamental research along the spectrum from curiosity-driven to use-inspired.
- Scanning for addressable fundamental issues of society and the economy; engaging with external counterparts to develop mutual understanding; and considering the potential impacts of the discoveries.
- Disseminating outcomes to the scholarly and partner communities to increase impact.
- Communicating with policy and funding bodies about the nature of impactful fundamental research and relevant timescales of impact.

The key outcomes are discoveries.

4.2.2 Engaging Researchers in Impactful Fundamental Research

The first key action in engaging researchers is to attract faculty to this practice, giving them freedom to undertake fundamental research along the spectrum, from curiosity-driven to use-inspired. Many will naturally adopt the curiosity-driven approach, but some may not fully realize the richness of use-inspired research.

The direct method is to share the arguments we present above, and let researchers reflect on the advantages of doing rewarding research that has both scholarly and societal impact. Evidence in the form of case studies can be effective, particularly if they include personal interactions with scholars who have adopted this approach, either from within the university or from an aspirational peer.

Nevertheless, there are cultural barriers to be overcome. A common barrier is suspicion by academics of becoming pawns of industry or government, or of wasting time on applied problems. We argue that, on the contrary, the approach enriches the whole research endeavor and can enhance the chance of unexpected breakthroughs.

Various forms of recognition and incentives can be created to support the faculty in this transition. Mentoring and professional seminars strengthen capacity. Expectations can be set to motivate this behavior. Measures related to use-inspired research can be included in the university's scheme of evaluation (Chapter 8). The case on the Australian National Universities references funding streams, assessment schemes, and the recognition system. All have been aligned to attract faculty to use-inspired efforts.

4.2.3   Identifying Opportunities for Use-Inspired Fundamental Research

Once there is interest on the part of the scholar, the issue shifts to finding a match between research interests and broader needs. This involves scanning for addressable issues of a fundamental nature in industry, enterprise, and government institutions, engaging with informed counterparts to develop mutual understanding, and giving consideration to the potential social and economic impacts of the discoveries.

There are various ways that a scholar can identify opportunities. Universities' strategic plans often address such issues, and the interested scholar can search within the university for those programs that are cited by the strategic plan or funded by its initiatives. The practice of Facilitating Dialog and Agreements creates a conversation in which scholars can participate. (Chapter 5).

The scholar can also engage directly with researchers and people in industry, enterprise, or government. These approaches include:

- Participating in national or regional discussions with thought leaders.
- Organizing visits by partners to the university, and by scholars to partners.
- Identifying the relevant partner point-of-contact for a university within a partner.
- Working in an entrepreneurial venture, or consulting with a partner.
- Developing proposals with partners for national programs.
- Working with government panels, institutions, and agencies.
- Reading the program goals of funding agencies and foundations.

4.2.4   Disseminating Outcomes in Academia and the Partner Community

The scholar should disseminate the results of both curiosity-driven and use-inspired research to the scholarly and partner communities so they can be understood by both. The scholarly outcomes will funnel through publications that might include a "theory paper" and an "application paper." The latter is often jointly authored with an industry partner. It may appear in a journal where the scholar does not usually publish.

Research discoveries are important to those seeking solutions to real challenges. The irony is that partners may not, at first, be ready to take on the discovery. It may appear risky, underdeveloped, or not fitting within existing technology programs. The practice on Maturing Discoveries and Creations is designed to help overcome this gap (Chapter 5).

But there is a larger point: research results may be useful beyond the circle of partner, collaborator, and sponsor [6]. It is therefore important to consider some form of wider and more active dissemination to industry, enterprise, and government by:

- Using the same networks for dissemination that were used to identify the need.
- Making visits to companies, agencies, or foundations.
- Publishing the application paper highlighted above.
- Arranging to be spotlighted in industrial journals or more general media.
- Patenting, which puts the discovery in highly searched databases.

4.2.5 *Communicating with Policy and Funding Agencies*

At the university level, there is another opportunity. Universities can communicate with policy and funding bodies about the benefits of impactful fundamental research, and relevant timescales of curiosity-driven and use-inspired outcomes. This communication can help build a bridge of understanding between the aspirations of most faculty, who tend to focus on the fundamental, and government and society who seek more direct impact from their investment in universities.

Good communication will support sound policy making. A better understanding of the benefits of a portfolio of impactful research along the spectrum from curiosity-driven to use-inspired can build support for both. Use-inspired fundamental research as part of a portfolio can satisfy an angst-ridden government concerned with return on investment. It can also broaden the perspective of funders to recognize the value of different approaches, motivations, and impacts.

4.3 Collaborative Research within and across Disciplines

4.3.1 *Motivation for Collaboration*

As a natural follow-on to Impactful Fundamental Research, the scholar will be exposed to challenges that are not neatly addressed within a single discipline. An appreciation will grow that, to address an issue, several disciplines or approaches may be necessary. This leads to the next practice: Collaborative Research Within and Across Disciplines. This complements, but does not replace, individual research efforts and funding. This practice focuses on collaboration among scholars at universities. The case of working with industry and enterprise is dealt with in the practices on Centres of Research, Education, and Innovation (below) and Maturing Discoveries and Creations (Chapter 5).

Box 4.4 Goals of Collaborative Research within and across Disciplines

Universities and their researchers will make discoveries
that are more effective instruments of knowledge exchange and innovation,
By collaborating with other scholarly researchers
within the university and externally,
Making discoveries that cross-disciplinary boundaries
and are in new fields of thought.

An important motivation for collaborative research is to make breakthrough discoveries at and across boundaries, which are more likely to lead to discoveries and creations with innovation impact. In cross-disciplinary collaboration, scholars bring together intellectual and real assets in order to better make discoveries at the edges of and across disciplinary boundaries (Fig. 4.3). Another motivation for collaboration is to link together the efforts of scholars with different methodological approaches. In this way, emerging ideas in new fields of thought can be translated to technologies that can more quickly have innovation impact.

Sometimes researchers collaborate just to secure more intellectual horsepower and gain access to unique assets: facilities, data sets, and funding. Collaboration is sometimes done to serve social or regional good [7]. Regardless of the motivation, researchers are still independent, but have yielded some control to colleagues.

In the past, when universities were small intimate communities with few organizational units, interdisciplinary discourse was informal and self-organized. The increase in student numbers and degree programs since the 1960s, and the

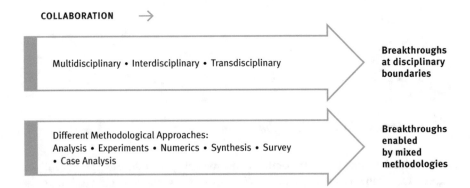

Fig. 4.3 Scholars engage in research collaboration by working across disciplines and with different methodologies

development of organizational units dedicated to disciplines, makes spontaneous cross-disciplinary discourse harder. University strategic plans generally highlight the importance of collaboration, but implementation is inconsistent.

Collaboration across disciplinary and organizational borders presents difficulties, and steps should be taken to make the work effective and efficient. There is evidence that the productivity of collaborators at multiple universities may be lower than if the same researchers worked within the same university. Working with collaborators adds a burden of coordination [8, 9]. In spite of these known challenges, many funding agencies explicitly encourage interdisciplinary and cross-institutional efforts. Because of these known systemic issues, we suggest that teams focus on collaboration style and process. If necessary, the university should provide facilitation.

Four key actions for implementing this practice are summarized in Box 4.5, and they are each discussed further below.

Box 4.5 Collaborative Research Within and Across Disciplines— Practice and Key Actions

Collaborative Research Within and Across Disciplines involves scholarly researchers from diverse fields, approaches, and sometimes institutions. They collaborate to make discoveries and exploit emerging ideas and technologies, often at the edges of existing disciplines and in new emerging fields of thought. They do this by exploiting the strengths and other assets of the participants.

Rationale: Breakthrough opportunities often exist at the boundaries of disciplines, by quickly connecting emerging discoveries and technologies, or through access to unique facilities or data sets. This often requires a scholar to voluntarily establish new collaborations and to recognize potential complementarities among researchers and institutions. The challenge is to work effectively and efficiently in such diverse teams.

Key actions:
- Collaborating across disciplines and approaches to bring in different methods and viewpoints.
- Collaborating around complementary resources, particularly large capital facilities, data sets, and funding.
- Setting high expectations on the value of outcomes, so that the benefits outweigh the difficulties of geographic distribution or institutional barriers.
- Conducting frequent in-depth interaction and exchange to facilitate collaboration and build relationships of knowledge and trust.

The key outcomes are discoveries.

4.3.2 Collaboration across Disciplines and Methodological Approaches

Academics routinely communicate with colleagues in their own disciplines, but less frequently across disciplines. Yet contemporary research often requires participants from diverse disciplines and institutions. Cross-disciplinary research has advantages, bringing together scholars from different disciplines, fields, or schools of thought.

Cross-disciplinary collaboration can be seen as a continuum characterized by [10]:

- Multidisciplinary collaboration: research is conducted by scholars from different fields, but addresses a question from one domain.
- Interdisciplinary collaboration: questions can span disciplines; scholars work together to formulate a problem and analyze and interpret research results.
- Transdisciplinary collaboration: this addresses problems that cannot be captured within existing disciplinary domains. Collaboration of this type can engender new "metadisciplines."

Said more simply, there are three schemes for collaboration: within a domain, at the boundary between domains, and those that lead to new emergent domains.

From the perspective of research for impact on knowledge exchange and innovation, we recognize that interdisciplinary and transdisciplinary collaboration are more risky, but are more likely to produce impactful outcomes. It is at the unexplored borders of disciplines that new and unexpected discoveries are most likely. This form of collaboration can lead to new scientific discoveries with potential for innovation (Fig. 4.3).

There is a second sense of collaborative research, involving scholars with different scientific and engineering methodological *approaches*. In such collaborations, experts in different approaches can be brought together: analytic theorists, numerical experts, model builders, experimentalists, synthesizers, etc. [11] Social science research methodology such as surveys and case studies can be useful. Creating this kind of collaboration has the potential to speed up the incorporation of ideas to from new fields of thought into outcomes that will accelerate knowledge exchange and innovation.

Collaborative research can sometimes be most exciting and productive when the collaborators come from dissimilar domains. This is illustrated by Case 4.3 from the Indian Institute of Technology Delhi on Interdisciplinary Collaboration in Public Policy for Science, Technology, and Innovation. Through the work of this flexible collaboration, scholars from technology and social science work together to influence public policy.

Case 4.3 Indian Institute of Technology Delhi (IITD)—Interdisciplinary Collaboration in Public Policy for Science, Technology, and Innovation

In a university with older formalized centers for Interdisciplinary research, newer more flexible academic collaborations bring together scholars from fields such as policy and technology, enabling them to influence policy and address societal challenges.

Interdisciplinary work at IITD began to flourish in the 1970s with the establishment of formal academic entities that were forward-looking for their time. They were led by visionaries who were empowered by hiring flexibility. For example, the Center for Atmospheric Sciences contributed significantly to the foundations of the Indian Meteorological Department, and the Department of Biochemical Engineering and Biotechnology influenced the development of the biotechnology sector in India.

Much of the research at these centers encountered difficulty in effectively translating to application, due to a lack of a broader enabling ecosystem. When the government itself was the main customer, the technology was more likely to be implemented. For example, the Center for Applied Research in Electronics developed the Phased Array Radar, a strategic imperative for the Indian defense establishment.

But over time, these entities became somewhat self-contained and inward looking. By the 1990s, interdisciplinary research at the Institute entered a period of stability, with no new centers established until the mid-2000s.

In recent years, the Institute has undergone significant rejuvenation in the establishment of effective interdisciplinary initiatives. For example, Faculty Interdisciplinary Research Projects give seed funding to professors from different academic units at IITD to engage in joint research. The School of Interdisciplinary Research facilitates the hiring of PhD students by faculty members from all areas of science and engineering. New entities (e.g., the Department of Design) are being created, and existing entities (e.g., the Center on Automotive Research and Tribology) are being transformed.

Another recent initiative is The School of Public Policy, which aims to be an academic center of excellence for domestic and global policy research, with a particular emphasis on science, technology, and innovation. It focuses on six broad overlapping topics: energy and environment; sustainable habitats; agriculture, food and water; health innovations and systems; industry and economy; and internet, digital information, and society.

The objectives of SPP are to: develop policy proposals, engage with high-level policy makers, promote a broader public dialog with stakeholders, and help build local capacity for policy analysis and implementation. SPP will enhance development of the next generation of scholars and practitioners through innovative academic programs (see Table 4.3).

The school takes a multidisciplinary approach to policy studies. In order to design technical and policy solutions that could protect and advance the public

Table 4.3 Academic programs envisioned by the School of Public Policy

Academic level	Distinctive features
Undergraduate	A minor with core courses designed in collaboration with various departments to enhance the general education of undergraduate students by ensuring that they have basic exposure to, and familiarity with, policy, legal, and societal aspects of science and technology issues
	Courses organized as seminars to provide the opportunity for students to engage in cutting-edge discussions on current public policy topics, e.g., Science, Technology, and Sustainable Development; Industrial Innovation and Organization; Health Systems and Innovations; Science, Technology, and the Future of Agriculture; and Information Infrastructure
Graduate	Master's degree in Public Policy (MPP) as well as Executive Education courses for policy makers and other senior personnel from relevant governmental, inter-governmental, and private organizations
	PhD program as well as a minor in Science, Technology, and Public Policy. Focus on areas such as Technology and Innovation Policy, Technology and Development, or Law and Technology (jointly taught with Humanities and Social Sciences and other departments)

interest, it is essential that the school hires faculty from a multiplicity of disciplinary backgrounds, such as engineering, physics, economics, sociology, and computer science. The school is also engaging with natural sciences, humanities, and social sciences colleagues from other parts of IITD.

The school will link with emerging IITD programs involving design, innovation, and entrepreneurship. It is also building linkages with a range of domestic policy entities, and international organizations such as the Red Cross, UNICEF, and UN Environment.

SPP aims to use its convening power to bring together domestic and global stakeholders, as well as engage with policy makers.

Based on inputs provided by Ambuj Sagar, Founding Head of the School of Public and Prof. M Balakrishnan, Deputy Director (Strategy & Planning) at the Indian Institute of Technology Delhi, Delhi, India.

4.3.3   Collaborative Research Based on Complementary Assets

An important rationale for collaboration is to gain access to valuable assets. It is essential to share major capital equipment and facilities (e.g., high energy physics facilities, spacecraft). These facilities can produce unexpected results. They stimulate joint planning. They can intrigue funders and university management with their promise. An expensive facility in this sense does not just provide a service, it serves as a platform for dialog. Collaboration naturally follows.

In the era of exponentially increasing data, scholars need access to good, large, open data sets. Likewise, access to numerical codes, computational power, and high bandwidth networks can be a significant stimulus for collaboration. Especially in the life sciences, access to new laboratory and experimental protocols can be an important benefit. Collaboration makes these assets more available.

In certain nations, collaboration is used to build research capacity—researchers from one institution can help to build research capacity at another. The availability of separate or additional funding also is often an important incentive for collaboration.

4.3.4   Expecting High Value Outcomes to Offset the Difficulties of Collaborative Research

Collaborating across disciplinary, institutional, and geographic boundaries is difficult. It can be less productive than work in small local disciplinary groups. The potential outcomes must be high value so that the perceived benefits outweigh the difficulties.

Even collaborating with peers in similar disciplines has challenges. There is some loss of independence, a need to share credit, and the problems of group dynamics. Working within the same university, but across disciplinary boundaries has the same problems, and it raises issues about terminology, conceptual constructs, and the different standards for the validation of evidence [10].

Working with another university requires dealing with variations in university culture, work ethic, incentive, intellectual property, and other regulation. These problems are compounded by working at a distance or with an international partner, surfacing issues of national law and tradition, and differing languages and time zones. Working with scholars outside the university community (e.g., those in national laboratories) adds another layer of questions involving intellectual property rights and student involvement.

Our observation is that the more "distant" the collaboration, the more difficulties will be encountered. To justify the difficulty, there must be greater expected benefit in terms of discoveries and impact on innovation. The difficulties can be eased if the collaboration begins with a formalized agreement guiding the participants in the resolution of issues including joint decision-making, group work norms, resource allocation, academic standards, intellectual property, communications, and conditions for termination. Ideally, this should be agreed as part of the proposal process and before work is begun.

4.3.5   Human Interaction as a Foundation of Effective Collaboration

Collaboration is fundamentally a process of human interaction. Knowledge of collaborators and mutual trust are key factor for success. "Researchers … desiring to work on interdisciplinary research, education and training projects should immerse themselves in the languages, cultures and knowledge of their collaborators [12]." "At the heart of interdisciplinarity is communication—the conversations, connections, and combinations that bring new insights to virtually every kind of scientist and engineer."

How does one create such collegial collaboration? Perhaps the most important element is a mutual intellectual interest between collaborators who want to learn from one another. It has been found that a prior project with a collaborator predicts greater strength of a current working ties [9]. The inference is that scholars should take every reasonable opportunity to take part in small joint projects or exploratory collaboration, progressively building links and preparing for larger future collaborations. In this way, there will be a pre-established level of knowledge and trust when the next opportunity for collaboration presents itself.

A scheme for building mutual interest into collaborative research was implemented at the University of Colorado, Boulder. As described in Case 4.4 on its Interdisciplinary Research Themes, a progressive process of faculty discussions, internal competition, and seed funding allowed a university with a more traditional disciplinary research organization to nucleate cross-disciplinary research themes.

Case 4.4 The University of Colorado Boulder (CU)—Interdisciplinary Research Themes

A modest set of incentives allowed an engineering college to organize collaborative teams, seek new resources, and accelerate contributions to interdisciplinary research.

The University of Colorado Boulder's College of Engineering and Applied Science (CU Engineering) has a strong tradition of research excellence within its departments, programs, and research centers. However, some of the most pressing technical and societal challenges are not within a discipline and require faculty and students to conduct research across fields in interdisciplinary teams. These interdisciplinary challenges necessitate a new approach to research collaboration, thus the college initiated six Interdisciplinary Research Themes (IRTs) [1] in January 2018:

- Autonomous Systems—smart, safe, and secure autonomy.
- Imaging Science—medical and seismic imaging, nondestructive testing, metrology.
- Multifunctional Materials—that mimic the integrated function of biological systems.
- Precision Biomaterials—integrated with drug and cell-based technologies.
- Quantum Integrated Sensors System—quantum science and technology for sensors.
- Water-Energy Nexus—challenges that also extend to food, land, air, and climate.

The IRTs are part of CU Engineering's overall strategy to build on its strengths, prepare for future opportunities, and accelerate its positive impact on Colorado and nation. Through the IRTs, the college plans to increase the number and size of funded research projects. The success of the IRTs will be used to guide investment of college resources, including faculty positions and shared research facilities.

CU Engineering established the IRTs through an open and competitive process. College faculty submitted 28 short IRT proposals in response to a request from the dean. These were reviewed by a group of senior faculty, and full proposals were invited from seven potential themes. Six IRTs were ultimately selected by a committee that included the department chairs, associate deans, and dean. The successful teams consist of self-selected college faculty and chose faculty directors to lead the efforts. Four of the six IRTs are led by early mid-career associate professors.

Each IRT director was provided with a budget of $750,000 over a 4-year period, for a total investment of $4.5 million. Metrics for success include industry collaborations, national reputation, and successful research proposals and projects. More than half of the internal IRT funding was awarded to faculty teams as seed grants to initiate promising research that could lead to larger funded opportunities. These IRTs are viewed as transformative collaborations, but not permanent organizations. After this initial seed investment, they are expected to become self-sustaining, and new themes will be identified and funded. Each IRT will be reviewed after 2 years.

Even in their first year, the IRTs have been a nexus of development. Significant new funding has been attracted, and interdisciplinary faculty search committees were created that recommended faculty appointments across these six areas (regardless of department affiliation).

Overall, research awards increased by 28% relative to the prior year, a record high level for the college. Additionally, the IRTs have catalyzed broad cross-campus partnerships and significant new engagement with national labs and industry partners. For example, the quantum IRT was originally established in the college, and within a year, it joined forces with a larger campus effort, the CUbit Quantum Initiative, to foster interdisciplinary work in quantum technology (see Fig. 4.4). As a result, the initial IRT funds have been matched four-fold by others.

Fig. 4.4 Quantum electrical and computing devices developed at CU Boulder

Contributed by Robert Braun, Dean, College of Engineering and Applied Science, The University of Colorado Boulder, Boulder, Colorado, USA.

Reference

1. (2018) Interdisciplinary research themes. In: College of Engineering & Applied Science. University of Colorado Boulder. https://www.colorado.edu/engineering/research/interdisciplinary-research-themes. Accessed 20 Jan 2020

What is needed is a reinvention and re-articulation of the old concept of collegial sharing and personal faculty engagement across disciplines. Such an effort would involve recruiting faculty who seek a barrier-less disciplinary environment, ensuring that barriers are low and providing recognition for cross-disciplinary action. Frequent communications, visits, and personnel exchanges contribute to effective collaboration. Immersing the whole team in the tasks at hand is important [10].

4.4   Centres of Research, Education, and Innovation

CENTRES

4.4.1   Directly Addressing Societal Issues

It is a logical progression from the practices of Impactful Fundamental Research and Collaborative Research to a stronger and more focused response to the significant challenges faced by industry and society. The practice of Centres of Research, Education, and Innovation (CREI) is about harnessing a critical mass of human, physical, informational, and financial resources to address these challenges at scale.

Box 4.6 Goals of Centres of Research, Education, and Innovation.

Universities and their researchers will make discoveries that are more effective instruments of knowledge exchange and innovation,
By empowering larger scale integrated Centres of Research, Education, and Innovation,
Producing directly implementable and impactful solutions that address significant issues of society.

Fig. 4.5 Centres of
Research, Education, and
Innovation producing
research outcomes,
impactful solutions, and
future leaders

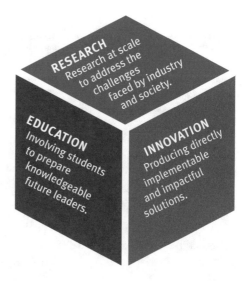

There is a sharp distinction between this practice and the two previous ones. Impactful Fundamental Research and Collaborative Research Within and Across Disciplines both focus on fundamental research by scholars.

This practice of Centres, however, calls for timely, direct, and implementable research, education, and innovation outcomes that impact society (Fig. 4.5). It unapologetically creates an integrated larger scale community of faculty plus stakeholder participants who might come from industry, enterprise, government institutions, national and regional governments, public interest groups, and elsewhere [13]. A CREI is normally led by a respected and effective thought leader with strong administrative support. These CREIs may be organized as informal collaborations, or as new structures that sit alongside or across existing organizations.

The practice calls for recruiting representatives of all the stakeholders early, so that the definition of the societal issue can be sharpened, its solution can be facilitated, and outcomes can be rapidly exchanged and moved towards applications. The involvement of stakeholder participants and students helps prepare a group of future leaders. The inclusion of other universities increases CREI's potential academic influence.

Such a Centre can be at a university or at an intermediary organization that is outside both the university and industry. Fraunhofer Institutes in Germany capture this role [14]. The organization decision is complex: closer to basic science and with less overhead of implementation favors the university setting; closer to the market and with access to industrial capability favors the intermediary. In a national innovation system, the two are complementary. We will focus on the university-based variant, where the university plays an important role by signaling the importance of the issue, recruiting faculty and students, and providing facilities.

Key actions for implementing this practice are summarized in Box 4.7, and they are each discussed further below.

Box 4.7 Centres of Research, Education, and Innovation—Practice and Key Actions

Centres of Research, Education, and Innovation are research communities that bring together scholars, experts from industry, government, and other groups. These participants advance the pace of knowledge exchange and the production of directly implementable and impactful solutions in research, education, and innovation that address significant issues in industry, economic development, and society.

Rationale: Involving stakeholders early better informs the framing of research questions and creates lower barriers to knowledge exchange. The focus on solving a problem of importance in society increases the potential impact of outcomes. The development of a community, the educational efforts, and the multidirectional exchange of people, practices, and ideas all help prepare a cadre of future leaders.

Key actions:
- Identifying the key issues to address, motivated by needs of society and overarching "grand challenge" goals.
- Building an integrated community of university researchers and students, industry, government, regulators, public interest groups, and others, working together in close partnership.
- Developing outcomes at an accelerated pace and testing them in realistic environments.
- Structuring an effective and well-defined organization to lead projects, manage interactions, and attract resources.

The key outcomes are discoveries, creations, and talented graduates.

4.4.2   Identifying Issues of Impact on Society

The issues that might be addressed by a Centre are similar to those in Impactful Fundamental Research. But now the scale of effort can be larger and the timescale more immediate. There are many actors available to identify issues that might be addressed:

- Learned societies that can identify grand challenges on the timescale of decades to centuries [15].
- Industry that can signal its needs for future technology [16].
- Think tanks, government foresight exercises, and announced long-term plans of funding agencies.
- Forecasts based on understanding of emerging science and learned fields of thought.
- Motivation by personal experience—congested cities, scarce resources, chronic disease.

In formulating a Centre, the founders may be responding to a call for proposals from a funding body. The university may have included the topic in its plans, or the university may be responding to a grassroots initiative. There is no shortage of important issues. The questions are: can the university frame a project with the right scope and a reasonable chance of impact, find an effective leader, raise the necessary money, and build a team with the right capabilities?

4.4.3   Building an Integrated Community

The next challenge in such an endeavor is building an integrated community to address the identified issue. This is the extension of the practice of Collaborative Research Within and Across Disciplines, but participants now explicitly include non-scholars.

The initiator of such a Centre will likely be a strong technical or scientific leader at the university. They will reach out to like-minded counterparts in industry, often trusted colleagues. It is desirable that one or several large companies be interested and that there be a proponent inside each company for the Centre. In this way, the advocates can help frame the vision of how a solution would be developed and transferred to impact. Lecuyer [17] credits such industry-university collaboration as a major force on economic development. Bringing in important partner participants early is important to their ownership of the solutions developed.

The combination of universities plus industry provides a platform for convening a wider group. First in this expanded list should be government. The roles of universities, industry, and government increasingly swirl together [18]. It is hard to imagine an issue of importance that will not involve regional or national government in some way, as a policy maker, funder, regulator, participant, implementer, customer, or competitor.

The convening power of the Centre can then be used to draw in other partner participants who have a substantial stake in the outcomes and who may be needed to assess social, economic, and policy ramifications (Engaging Stakeholders in Chapter 7):

- NGOs, public interest groups, and policy bodies.
- Operators of the system and labor.
- Media.
- Small and medium enterprises that have been identified by the larger industries as important suppliers.
- Government institutions (e.g., national laboratories, institutes of the scientific academy).
- Local communities.
- Sources of capital and investors.

These partner participants may be important bridges to communities that will participate in implementing from a solution and will benefit from it. These participants may be organized in tiers: some may be strategically involved and contribute major funds to the Centre; others may be less deeply involved, taking on the outputs of the Centre.

An example of a Centre can be found in Case 4.5 on the CMI Knowledge Integration Communities (KICs). An extraordinary social-technical problem—airport noise—was identified, and a particularly broad team of engineers, operators, and regulators was formed. Involvement of diverse participants sped up computational design. Rapid testing led to early results, and the identification of a new concept for the "silent aircraft."

Case 4.5 The Cambridge—MIT Institute (CMI)—Knowledge Integration Communities (KICs)

Joined up communities of scientists, engineers, manufacturers, operators, and regulators can address concrete problems of society, while also producing valuable research, innovation, and educational outcomes.

Airports are drivers of regional economic development, but their creation and expansion is often limited by local residents' legitimate concerns regarding noise. Universities should be willing to take on this kind of technical problem that has broad social impact. But we realized that the linear, or Frascati model (basic research, applied research, development and manufacture) is inadequate to deal with these complex real-world problems.

During the operation of the Cambridge—MIT Institute (2000–2007), we created an alternative model known as the Knowledge Integration Communities (KIC) [1]. These KICs employed a kind of spiral development model where the spirals stretched to include fundamental research. We created six KICs that were:

- Focused on a societal scale issues addressable by science, technology, and policy.
- Assembled to include the stakeholders from basic research through to agents of widespread deployment.
- Encouraged to have close and open interaction, and to view and solve the issues in an integrated manner.
- Expected to produce outcomes that would have direct benefit.

The Silent Aircraft Initiative was an example of what we sought to achieve. The *focus* was to design an aircraft which would be functionally quiet—not heard above the urban background noise outside the boundary of the airport. This would significantly reduce the community concerns with airport development.

The engaged stakeholder community had strong collaborations with other universities, national labs and with manufactures. The real advance was to include other important parties, especially the operators of the aircraft and airports: airport authorities, passenger and cargo airlines, air traffic controllers, and pilot organizations (see Fig. 4.6). Various government agencies, regulators, and investors were also involved. There was also significant outreach to agents of formal and informal education.

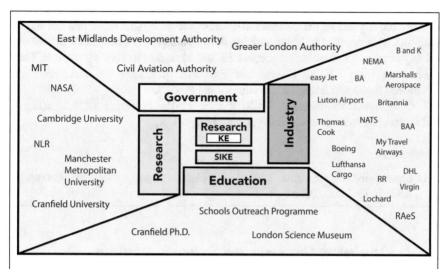

Fig. 4.6 The participants in the Silent Aircraft Initiative KIC

Fig. 4.7 The SAX-40 Silent Aircraft

The nature of the *close and open interaction* was remarkable. Boeing and Rolls Royce gave access to their proprietary design software, so that time was not spent redeveloping code. The regional airports offered landing slots for trial flights with new abatement measures. The freight airlines were targeted as first users of such an aircraft, which might require new flight maneuvres.

The technical *outcome* was a flying wing that omitted many of the smaller lifting surfaces that increase lift but generate noise. Takeoff noise was reduced by using a larger diameter engine with higher thrust and lower noise, and placing it above the body so that the body would block downward noise radiation. The landing noise is reduced by flying a shorter, steeper descent controlled by drag created by the engines (see Fig. 4.7).

Equally interesting was the result of an economic exercise to estimate the real cost of noise, which, when applied to Heathrow Airport, which, if implemented, could imply an uplift of local property values in the range of billions of GBPs.

Within 2 years of the project's 2006 completion, NASA identified the silent aircraft as a model for future aircraft development. The European Institute of Innovation and Technology incorporated KIC ideas into its funding streams. The evolution from KIC to Centres for Research Education and Innovation at Skoltech involved adding formal educational aspects.

Contributed by Michael Kelly, Prince Philip Professor of Technology Emeritus, University of Cambridge, Cambridge, UK.

Reference

1. Cambrige-MIT Institute (2008) Accelerating Innovation by Crossing Boundaries 2000–2006.

4.4.4   Developing Outcomes and Testing in Realistic Environments

The next step is to develop a vision and roadmap for the project. It should be emphasized that this is a notional plan, which can be amended by the team (Mission and Strategic Planning in Chapter 7).

An effective way to engage all participants in the main issues is to work on a representative sample problem or testbed. By engaging participants early on a representative problem, they tend to become more aware of where progress needs to be made.

The outcomes of research will be new discoveries and creations. For Centres, these will take the form of integrated designs or solutions. One valuable contribution of the Centre is to test these solutions under realistic conditions. The actual implementation of this idea is highly variable—from medical trials, to the operations of a new procedure at an airport. But test in these conditions will attract interest, demonstrate effectiveness, and highlight open issues for further work.

These tests in realistic environments are part of well-organized efforts to exchange knowledge effectively with partners. There should be integrated knowledge exchange plans and mechanisms, with well-defined organizations to manage interactions and projects [19] (Maturing Discoveries and Creations and Facilitating Dialog and Agreements both in Chapter 5).

Not to be overlooked is the role of Centres in education. The new ideas and discoveries that are made should quickly migrate to graduate education (Education in Emerging Thought in Chapter 3). Students of course will be engaged in the research of the Centres (see below).

Such Centres can be small or large, informal or formal. An example of a fairly large but informal Centre is the Antimicrobial Research Collaborative (ARC) at Imperial College London (Case 4.6). The ARC addresses an important issue for healthcare systems and society using a broad multidisciplinary approach, engaging scholars from physicians and pharmacologists to mathematicians and policy makers.

Case 4.6 Imperial College London—Antimicrobial Research Collaborative

Broad societal challenges can be addressed effectively by informal interdisciplinary collaborations that influence practice, policy, science, and education.

Imperial College London encourages collaboration across disciplinary and organizational boundaries in order to impact important issues of society. One such contemporary global issue in healthcare is increasing antimicrobial resistance.

Antimicrobial resistance (AMR) occurs when microorganisms are able to survive exposure to antimicrobial medicines such as antibiotics that would normally kill them or stop their growth. This results in the drugs no longer able to treat infections. Such super-bug infections, which were once easily treatable, can become fatal. AMR is accelerated by the inappropriate use of antimicrobial medicines, incorrect prescribing, and poor infection control practices.

The Antimicrobial Research Collaborative addresses these issues with broad, multidisciplinary approaches [1]. ARC@Imperial harnesses the expertise of Imperial's leading physicians, molecular bacteriologists, clinical pharmacologists, bioengineers, clinical trials experts, chemists, primary care specialists, allied health professionals, computational biologists, mathematical modelers, statisticians, environmental experts, policy makers, business analysts, evolutionary biologists and epidemiologists, working alongside numerous external partners.

The collaborative involves more than 100 academic experts working across the faculties of medicine, natural sciences, engineering, and the business school at Imperial College London and Imperial College Healthcare NHS Trust.

The collaborative is not a formal organization and was largely formed by a self-assembled group of those willing to collaborate. There are no obligations on the participants. The collaborative works because the participants find benefit. In addition to informal networking, there is an annual meeting where collaborators summarize findings. Other coordinating meetings and workshops are set to address cross-cutting issues. The collaboration and its activities also provide clear evidence of experience and the potential for public engagement and impact that funding bodies increasingly seek in research proposals.

While there are small amounts of funding available to facilitate interaction, a more effective means of interaction in the early days of the ARC was the use of young fellows. About ten fellows from diverse backgrounds played an important role building the community. Because each had a primary group affiliation and at least one additional affiliation, they helped build bridges across groups.

The ARC collaborators are having an impact on research, education, policy, and translational medicine. For example, researchers across medicine, chemistry, and bioengineering are developing wearable biosensors to monitor antibiotic levels and individual patient responses, so that dosing of antibiotics can be optimized and dynamically individualized. Integrated innovative point-of-care diagnostics have been developed, with connectivity embedded, so that epidemiology and public health interventions can be considered. The collaborations are also facilitating the necessary codesign required for technological advances, including the application of data sciences and artificial intelligence.

Educational efforts build capacity in crossing domains and working together, which is especially important for younger scholars. An area of important work by ARC is its contribution to policy governing use of antibiotics, as well as the associated clinical practice. The partnership of Imperial College with the Imperial College Healthcare NHS Trust is a pathway for translating discoveries into medical advances, new therapies, and techniques, while promoting their application in the NHS and around the world.

Prepared based on input provided by Alison Holmes, Professor of Infectious Diseases, Imperial College London, London, UK.

Reference

1. The Antimicrobial Research Collaborative. In: Imperial College London. https://www.imperial.ac.uk/arc/. Accessed 20 Jan 2020

4.4.5   Structuring and Managing the Centre

Managing the Centre requires both thought leadership and skill at community building, managing interactions, and administration. It is best accomplished by a team composed of a faculty director working with, perhaps, a faculty codirector, an executive director from industry or both.

Issues of governance should be resolved early in a "constitutional document" of the Centre. In addition to outlining the decision-making process, this document should describe the scope of the Centre, its governance, and how the Centre will be resourced. The constitution should specify the degree of independence of the Centre and how it fits with the other structures of the university. The document should explain how the research of the Centre informs education and catalyzing innovation, discussed in Chapters 3 and 5, respectively.

The executive director would be responsible for planning, communications, industrial outreach, advocacy, and stakeholder engagement. They help shape and monitor the budget, human resources, and reporting. In addition, the executive director works within the university to develop contracts and legal agreements. Principal among these are intellectual property agreements. The executive director also arranges facilities and plays a role in developing resources. If the Centre is to be successful, it must achieve funding at a scale to reach critical mass. Finally, the executive director has an important role in community building, which will take significant time and effort.

4.5 Undergraduate and Postgraduate Student Researchers

4.5.1 Reinventing the Apprentice Model for Student Researchers

"The competitive advantage of the university, over other knowledge-producing institutions, is its students [18]." They are central to the university. They bring in new ideas, they work hard, and they learn a great deal. Desirably, they leave with knowledge that can influence economic development and with the will and self-confidence to have an impact.

> **Box 4.8 Goals of Undergraduate and Postgraduate Student Researchers**
> Universities and their researchers will make discoveries
> that are more effective instruments of knowledge exchange and innovation,
> **By engaging undergraduate and postgraduate students in research,**
> *Energizing research and yielding effective young researchers*
> *and agents of knowledge exchange.*

Involving students in research is part of their formulation and education. Research helps them in several distinct ways. It trains them in the process of research and its ethics and responsibilities. When properly framed, the relationship that develops with a research supervisor can be one of the longest lasting and important contributions to their education. The discovery of new knowledge through research complements the acquisition of knowledge in education, helping to build deep understanding and self-efficacy. Some students will use the specific preparation to build careers in research.

Student researchers are significant contributors. They bring imagination, energy, depth of specific knowledge. and scholarly visibility into literature. Many have advanced knowledge of emerging fields (Education in Emerging Thought in Chapter 3). They invigorate the research process, both because of their contributions, and by the constant renewal of energetic participants. Every student researcher can contribute to a progression of activities and responsibilities, along what we call the *intellectual continuum*: from first year students to experienced undergraduate researchers, on to master's, PhDs, and postdoctoral fellows. Some continue to be faculty.

In a university with a mission that includes economic development, students take on additional roles. They are arguably the most effective agents of knowledge exchange. Even during their student years, they can work with partners and contribute their knowledge. Then they graduate with a facility in cross-disciplinary and emerging thought, and a knowledge of discoveries and creations. They gain the skills to evolve as stronger researchers and innovators. They contribute in many ways throughout their career. Ideally industry, enterprise and government institutions have the wisdom and absorptive capacity to make the best use of these students in the workforce. Even while at the university, students can assess the innovation potential of discoveries, and alert colleagues to this potential.

Four key actions for implementing this practice are summarized in Box 4.9, and they are each discussed further below.

Box 4.9 Undergraduate and Postgraduate Student Researchers—Practice and Key Actions

Undergraduate and Postgraduate Student Researchers is the process of mentoring research students by involving them in a progression of research responsibilities, including work with partners and the assessment of the innovation potential of the research. The students will become effective agents of knowledge exchange and researchers who invigorate discovery.

Rationale: Research empowers the students through application of newfound knowledge, building of deep working understanding and increasing self-efficacy. Student researchers strengthen research, contributing enthusiasm, creativity and contemporary knowledge and tools. When they leave the university, student researchers can become a prime channel of knowledge exchange. They may become future independent researchers.

Key actions:

- Engaging postgraduate students in the mainstream of research, sometimes with partners, mentoring their development, and recognizing their contributions to publications and in scientific forums.
- Involving undergraduates in research partnerships with faculty as part of the faculty's research, or in student proposed research and innovation activities.
- Systematically including in each student's thesis an assessment of the innovation potential of the emerging research discoveries.

The key outcomes are discoveries, as well as the identification of their innovation potential, and the education of talented graduates.

4.5.2 Postgraduate Researchers

In postgraduate research, students take part in the mainstream of scholarly research. They participate in research groups, contributing substantially to discoveries and to publications and scholarly presentations. These activities prepare them to be independent technical and scientific leaders. They may spend time at the site of a partner, better understanding their needs, and exchanging knowledge with the partner.

In former times when there were fewer student researchers, they followed an apprentice model, launching into the deep end of active research from the beginning, and learning by following the practices of the supervisor. This was a reasonable approach to the training of future faculty in a time when students could reach the state of the art without formal coursework.

The explosion in the numbers of postgraduate students is partly due to their growing importance to nonacademic sectors. Students now normally participate in formal coursework, in addition to academic supervision. But even this is not sufficient. We seek a broader preparation—the new apprenticeship—that ideally includes:

(a) Strong mentoring by qualified supervisors to guide the student.
(b) Introductory coursework to equip the student with state-of-the-art knowledge in their chosen area and in acceptable research practice.
(c) A substantial research project following the apprentice model.
(d) Provision of co-advisors from other relevant disciplines and from industry, to expose the student to different norms of behavior and contextualize the discoveries.
(e) The development of skills beyond those of doing research, including aspects of entrepreneurship, leadership, and management (Preparing for Innovation, Chapter 3).

In the case of master's students, there is considerable variation from purely taught programs to intense research experiences. To the extent practical and allowed, a master's program can apply these elements of the new apprenticeship, but within shorter timescales.

As the competition for junior faculty positions has intensified, and the research process has become more complex, the number of postdoctoral researchers has also grown. Postdoctoral programs have some elements of PhD preparation (e.g., mentorship, publications), but should add some elements of professional development, as discussed in the supporting practice on Faculty and Staff Resources and Capabilities (Chapter 7).

The case of TUM—PhD students in industry exemplifies an aspect of the practice of Student Researchers (Case 4.7). Under this program, PhD students actually work in partner organizations and locations, usually with both a local partner supervisor and a faculty supervisor. Student and faculty more deeply learn the needs of the partner and build pathways for knowledge exchange, eventually catalyzing innovation.

Case 4.7 Technical University of Munich (TUM)—PhD Students in Industry

PhD students who perform their research work in industry build bridges of knowledge between the university and industry and help to catalyze innovation.

To speed up innovation, TUM partners with industry to transmit knowledge and inventions from academia to enterprises. TUM promotes both basic research and applied research, which focuses on concrete solutions to defined problems. TUM staff help industry to solve highly complex technological challenges, and this research is valuable to companies across the world.

TUM has played a crucial role in Bavaria's transition from an agrarian state to one of the leading high-tech hubs in Europe. This was due in part to strong government investment in strategic technological sectors, alongside government-facilitated collaboration among industry, universities, and the public sector [1]. This was organized around industrial sectors, such as aerospace, automotive, high-tech agriculture, and electrical equipment. As a result, TUM has cultivated long-standing partnerships with global giants such as Siemens and BMW.

TUM has also forged alliances with research institutes; in particular, Max Planck Institutes, the Helmholtz Zentrum München, and the Fraunhofer Society. TUM has complemented this approach with programs to support new business creation, for example, in its entrepreneurship arm, UnternehmerTUM.

Doctoral students play an important role in this network that supports innovation. TUM has developed several distinct doctoral paths that maximize student learning and optimize societal impact. Besides the traditional university-based PhD, in which students are an academic researcher, several alternative doctorate paths exist, at TUM and elsewhere, that are frequently seen as a key factor in Germany's competitive advantage in industrial research and development.

In the *industry-based PhD track,* students are employed by a partner company or public sector organization. Students do research mainly at the partner organization under a company advisor, but they are also integrated into the academic environment at TUM through a dissertation supervisor. This results in a smooth flow of ideas between the university and industry because the student works at both places. What evolves is not necessarily a develop and transfer model, but more of co-development and mutual inspiration model. An external doctorate is often very hands-on, and lets students become part of the industry at an early stage. Industrial and other external PhD students are typically funded by the external partner and do not have any teaching requirements.

Thanks to strong focus on applicability and academic rigor, the TUM PhD students are typically received favorably by Barvarian industry. They form a bridge between the university and industry. They transfer knowledge to industry and support innovation. They grow and become educated and highly employable young professionals, often taking up employment in the local labor market.

Another PhD path is the *cooperative doctorate*. Currently in Bavaria, universities of applied sciences do not have the right to award doctorates. However, there are many excellent young scholars at these institutions who seek one. To harness the strengths of all of these universities, a joint framework for cooperative doctorates was developed under the auspices of the Bavarian Academic Forum (BayWiss). In this model, doctoral candidates are supervised jointly by one professor from TUM and one from the university of applied sciences. Once candidates successfully complete doctoral research, TUM awards the doctorate.

Prepared with input from Sebastian M. Pfotenhauer, Carl von Linde Assistant Professor of Innovation Research, Munich Center for Technology in Society, and TUM School of Management, Technical University of Munich, Munich, Germany.

Reference

1. Meyer-Krahmer F (2002) The German innovation system. In: Laredo P, Mustar P (eds) Research and innovation policies in the new global economy: an international comparative analysis (new horizons in the economics of innovation). Edward Elgar Pub, Cheltenham, UK, pp. 205–252

4.5.3 Undergraduate Researchers

In undergraduate research projects, sometimes called Undergraduate Research Opportunities, students work with faculty on the mainstream of the faculties' research. The participation is co-curricular and voluntary. Students commonly work as assistants in the laboratory on a wide variety of authentic research tasks. Alternatively, students can propose research or innovation activities that are self-organized and directed, but involve faculty supervision and mentoring. The discussion can be fuzzy since students often branch off in directions stimulated by the professor's work. These projects may be distinct from undergraduate theses because of the authentic and voluntary nature of the contribution.

The experience of being engaged in actual research empowers students to understand the relevance of classroom learning and to apply this knowledge. It builds deep understanding and increases self-efficacy. Students experience actual moments of discovery, which builds excitement and interest. Because of their exposure to research results, even students at this early stage of their career become agents of knowledge exchange. Many are stimulated to follow advanced degrees, or careers in science, engineering, and technology [20].

Good practice in Undergraduate Research includes:

(a) Recognition that student participation in faculty or independent research is important to the undergraduate learning experience.
(b) Student participation in authentic aspects of the research process, including proposal preparation and planning, designing and building equipment, coding,

conducting experiments, and analyzing data, as well as written and oral presentations.

(c) Student-identified research or innovation projects that involve meaningful faculty mentoring and supervision.

(d) Program options that allow the students to earn academic credit or compensation.

Undergraduate students' involvement in research programs varies. Dozens of universities have undergraduate research programs. In university programs in science and engineering, voluntary participation rates can range up to 80% of the eligible students [20].

A good example of including undergraduates in research is discussed in the case HKUST—Undergraduate Research Opportunities Program (Case 4.8). Students are engaged in authentic research under the supervision of a faculty member, providing context for learning of fundamentals and participation in the excitement of discovery. The students contribute to research outcomes, help solve problems, and carry research knowledge with them when they graduate.

Case 4.8 The Hong Kong University of Science and Technology (HKUST)—Undergraduate Research Opportunities Program (UROP)

Engaging in faculty-mentored research projects helps students to nurture curiosity and develop practical skills, while providing opportunities for them to cogenerate knowledge, solve real-world problems, and transfer knowledge beyond the university.

In 2005, The Hong Kong University of Science and Technology launched the Undergraduate Research Opportunities Program (UROP), providing students with an exciting opportunity to engage in academic research. This helps them to develop insightful perspectives on their areas of interest and to advance the frontiers of knowledge. UROP allows students to immerse themselves in tailor-made research projects across disciplines, under the supervision of world-class researchers. Students explore under the guidance of experienced researchers, leading them along the path of new discoveries.

The research opportunities in UROP fall into two streams—the Tasting Stream and the Series Steam—to meet the needs of students. The Tasting Stream is suitable for students who would like to get a first taste of the research experience. Those students that successfully complete the stream will be nominated for a modest stipend.

The Series Stream is designed for students who develop a serious commitment to research on the basis of their initial experience in the Tasting Stream. The Series Stream comprises four courses that must be taken in sequential order and under the supervision of the same faculty member for the same project although it is not necessary for them to be taken in consecutive semesters. The supervisor must consent to continue the course series.

Students are restricted to one UROP project per semester and are expected to devote at least 3 h/week to laboratory work on any given research project although the actual level of commitment is mutually agreed upon by the supervising faculty member and the student. Each faculty member is permitted to supervise a maximum of five projects and no more than ten students in any single term. Furthermore, supervisors must arrange a minimum of one contact hour per week for each project.

The UROP sponsorship scheme is intended to give UROP students financial support to publish their papers in international journals, to present their posters or papers at academic conferences, or to participate in research-related summer schools or workshops. Each year, abstracts from UROP students' research reports are compiled in the UROP Proceedings to showcase students' research achievements.

To encourage high-quality research, students who exhibit excellent research performance are eligible to receive the Mr. Armin and Mrs. Lillian Kitchell Undergraduate Research Award. The winning students' supervising team is presented with the UROP Faculty Research Award.

The International Research Opportunities Program (IROP) is an initiative under the auspices of UROP to broaden HKUST students' learning experience by giving them an opportunity to conduct research in partner overseas universities. The IROP experience helps students to enhance their technical and communication skills, while simultaneously promoting greater awareness and understanding of different cultures. The hosting university provides free on-campus accommodation for IROP students, offering the same on-campus access and privileges as that provided to local students.

UROP has been well received by both students and faculty, and enrollment has reached more than 450 students in the 2017–18 academic year, increasing every year since its establishment. Students benefit from UROP, building research and project skills, and gaining insight about their next professional step towards work or continued education. Faculty benefit from the program through the interactions with curious and creative millennials, who often bring unconventional and unexpected ideas to the project. And, the university benefits when graduating students carry their UROP acquired knowledge to a career position.

Prepared based on input provided by The Hong Kong University of Science and Technology, Hong Kong, China.

4.5.4 Student Assessment of the Innovation Potential of Research

Students are in a good position to assess the innovation potential of the research project on which they are working. They are knowledgeable about the content of the research, many are interested in innovation, and they have access to relevant faculty and staff, as well as online resources.

This assessment can be done by including in each student thesis or report an explicit section that assesses the innovation potential of the research outcomes. This can be included at any level of student research, from undergraduate to postdoctoral. If all students include this, there will be a systematic review of research to identify emerging discoveries and technologies with higher likelihood of innovation impact. All that is suggested is an assessment; a conclusion that the research result does not have strong innovation potential is acceptable. The direct outcome of this activity is a regular comprehensive review of all university research projects for innovation potential.

There are other benefits as well. Requiring an innovation assessment of all students' research projects will focus students on the innovation value of their work (as well as its social or cultural impact). If all students are required to do this, then all supervisors will also be drawn into considering the innovation potential. Desirably, there are staff and partners who can serve as *innovation advisors*, helping the student with this assessment (Faculty and Staff Resources and Capabilities Chapter 7 and Facilitating Dialog and Agreements Chapter 5). Collaboration of the student, research supervisor, and innovation advisor creates a pathway to exploit the results.

4.6   Chapter Summary

We present four practices that allow researchers to create new knowledge and discoveries (Fig. 4.8). These practices also provide for knowledge exchange that creates the potential for accelerating innovation and entrepreneurship. They emphasize the systematic practices of research, which are closely integrated with education

IMPACTFUL FUNDAMENTAL RESEARCH	• Pursuing curiosity-drive and use-inspired discoveries. • New knowledge with impact on scholars and society.
COLLABORATIVE RESEARCH	• Collaborating with internal and external scholars. • Discoveries across disciplines and in new fields of thought.
CENTRES	• Empowering Centres of Research, Education & Innovation. • Directly implementable and impactful solutions.
STUDENT RESEARCHERS	• Engaging undergraduate and postgraduate researchers. • Preparing researchers and agents of knowledge exchange.

Fig. 4.8 Research practices, processes, and outcomes

and catalyzing innovation, and exemplify the principle of systematic knowledge exchange (Box 2.3).

Impactful Fundamental Research is aimed at increasing its impact by engaging scholars in fundamental research along the spectrum, from curiosity-driven to use-inspired. The important criterion for good research is impact, and though our lens, impact on economic development. The university should enable scholars to learn about the needs of society and partners, and more effectively capture the value of fundamental research all along the spectrum. As the foundational practice in research, it underlies the other research practices.

Collaborative Research Within and Across Disciplines is a practice that brings together university scholars from various backgrounds and approaches to address issues of wider scope and more cross-disciplinary nature than can be done by an individual. It follows on from the practice of Impactful Fundamental Research and supports the practices of Centres of Research, Education, and Innovation and Maturing Discoveries and Creations (Chapter 5).

Centres of Research, Education, and Innovation assemble efforts at scale and seek directly implementable and impactful solutions that rely on research at the frontier of knowledge and address significant issues in society. The primary outcomes are discoveries. But the practice lies at the nexus of research, education, and innovation, so other important outcomes are innovation creations and well educated research students.

Undergraduate and Postgraduate Student Researchers aims to invigorate research with the curiosity of students, to train young researchers and support direct diffusion of knowledge. The primary outcomes are discoveries, but the practice also sits at the intersection of research, education, and innovation, so other important outcomes are innovation creations, educated students, and the assessment of the innovation potential of the research.

Not surprisingly, there is a good correspondence between these four practices and the policies of funding agencies. For example, the European Research Council (ERC) programs called Starting, Consolidator, and Advanced grants look a lot like Impactful Fundamental Research, emphasizing the curiosity-driven end of the spectrum. The ERC also sponsors add-on funding called Synergy grants; these support part of what we call Collaborative Research Within and Across Disciplines. ERC's Proof of Concept grants are a start at funding the activities included in Maturing Discoveries and Creations (Chapter 5). The US National Science Foundation (NSF) also sponsors a program of centers that are of scale and call for a route to near-term impact. These are much like our Centres of Research, Education, and Innovation.

As a final note, we reflect on the impact of these practices on sustainability in economic development. The investment in research will have near-, mid-, and long-term returns in sustainable innovation and entrepreneurship. The topics of the research program can be influenced to reflect sustainability of the economy and our world. Finally, the way in which the research is done can sometimes be designed to be more beneficial to the environment, for example, by design of experiments and facilities.

References

1. Cambrige-MIT Institute (2008) Accelerating innovation by crossing boundaries 2000–2006
2. Stokes DE (1997) Pasteur's quadrant: basic science and technological innovation. Brookings Institution Press, Washington, DC
3. Brooks H (1968) The government of science. MIT Press, Cambridge, MA
4. Conant JB (1951) Science and common sense, 16th edn. Yale University Press, New Heaven, CT
5. Conant JB (1947) On understanding science. Yale University Press, New Heaven, CT
6. Owen-Smith J (2018) Research universities and the public good: discovery for an uncertain future. Stanford Business Books, Palo Alto, CA
7. Dowling DA (2005) The Dowling review of business-university research collaborations. Royal Academy of Engineering
8. Cummings JN, Kiesler S (2005) Collaborative research across disciplinary and organizational boundaries. Soc Stud Sci 35:703–722
9. Cummings JN, Kiesler S (2008) Who collaborates successfully?: Prior experience reduces collaboration barriers in distributed interdisciplinary research. In: Proceedings of the 2008 ACM Conference on Computer Eupported Cooperative Work. San Diego, CA, pp 437–446
10. Eigenbrode SD, O'Rourke M, Wulfhorst JD, Althoff DM, Goldberg CS, Merril K, Morse W, Nielsen-Pincus M, Stephens J, Winowiecki L, Bosque-Pérez NA (2007) Employing philosophical dialogue in collaborative science. Bioscience 57:55–64
11. Galison P (1997) Image and logic: a material culture of microphysics. University of Chicago Press, Chicago, IL
12. National Academy of Sciences National Academy of Engineering and Institute of Medicine (2005) Facilitating interdisciplinary research. The National Academies Press, Washington, DC
13. Ackworth EB (2008) University–industry engagement: the formation of the knowledge integration community (KIC) model at the Cambridge-MIT Institute. Res Policy 37:1241–1254
14. Fraunhofer-Gesellschaft (2019) 70 years of Fraunhofer: 70 years of future. Munich, Germany. https://www.fraunhofer.de/content/dam/zv/en/Publications/Annual-Report/2018/Fraunhofer-Annual-Report-2018.pdf. Accessed 2 Jan 2020
15. National Academy of Engineering (2017) NAE grand challenges for engineering, Washington, DC
16. Fontana R, Geuna A, Matt M (2006) Factors affecting university-industry R&D projects: the importance of searching, screening and signalling. Res Policy 35:309–323
17. Lécuyer C (2005) What do universities really owe industry? The case of solid state electronics at Stanford. Minerva 43:51–71
18. Etzkowitz H (2008) The triple helix: university—industry—government in action. Routledge, London
19. Bloedon RV, Stokes D (1994) Making university/industry collaborative research succeed. Res Technol Manage 37:44–48
20. Russell SH, Hancock MP, McCullough J (2007) Benefits of undergraduate research experiences. Science 316:548–549

Chapter 5
Catalyzing Innovation and Knowledge Exchange

5.1 Introduction

5.1.1 The Objectives of Catalyzing Innovation

Actual innovation—the development and delivery of new goods, services and systems, and their associated operations—is primarily done by partners in industry, small and medium enterprise, and government organizations. Universities enhance this actual innovation by *catalyzing innovation* within the university. This involves synthesizing *creations*—artifacts and procedures that have never before existed and exchanging knowledge of these with partners. Improving the pace and effectiveness

---

Box 5.1 Objectives of Catalyzing Innovation and Knowledge Exchange

The general objective of catalyzing innovation is to produce *creations*—synthesized objects, processes, and systems that have never existed prior to their development at the university, and that have potential for societal impact. Creations include technologies, inventions, and other intellectual property, artifacts, methods and concepts, tangible research property (e.g., engineering prototypes, drawings, new organisms, software, circuit chips), know-how, and business ideas. But the creations of a university also include all things synthesized at the university, such as medical procedures, urban plans, and works of art (Fig. 2.1).

These can be exchanged by the same mechanisms associated with research: publications, discussions, joint projects, and personnel exchange. Additional mechanisms include intellectual property and tangible research property agreements, exchange of tangible artifacts, and involvement in start-ups and consulting. In addition, there are events, shows, and various networks. Graduating students are also a primary pathway for exchanging knowledge of creations (Fig. 2.4).

© Springer Nature Switzerland AG 2020
E. Crawley et al., *Universities as Engines of Economic Development*,
https://doi.org/10.1007/978-3-030-47549-9_5

of developing and exchanging creations is fundamental to accelerating innovation and producing economic gains.

Viewed through the lens of economic development, the specific objective of catalyzing innovation is to *more effectively stimulate and capture the rich innovation creations of a university and exchange them with partners in industry, small and medium enterprise, and government organizations.*

Referencing the discussion of Chapter 2, we remind ourselves that the university contributes to economic development by *accelerating innovation* and entrepreneurship. This is done by *enhancing knowledge exchange* between the university and its partners. Knowledge exchange is strengthened when there is a *systematic approach* that carefully identifies needs of the partners and society, conducts university activities with a sensitivity to these needs, and proactively exchanges the university outcomes with partners. The knowledge that is exchanged flows from the university's *cross-disciplinary and integrated activities,* including education, research, and catalyzing innovation.

5.1.2   Background and Opportunity

We now turn to the third of these interconnected domains, catalyzing innovation, and its associated knowledge exchange. While most universities are active in innovation, many view catalyzing innovation as a top-down function of a specific office, which often prioritizes revenue generation through patenting, licensing, and spinoffs [1].

Catalyzing innovation is a much broader endeavor. No single pattern of activity is best. We propose a broad array of top-down and bottom-up actions that the university could adapt to local conditions. Partners should engage with universities along a range, from more ad hoc approaches dealing with individuals, to more strategic approaches that identify important universities and tackle more ambitious open-ended challenges [2].

To be more effective, catalyzing innovation must be fully integrated into the university, not merely a bolt-on addition. Typically, the process of catalyzing innovation faces several problems. First, there is a considerable readiness gap between the maturity of creations developed at universities and the relatively higher readiness expectation of partners. Second, there is a need for multifaceted informal and formal interaction between universities and partners. Finally, there is a need to better prepare and experienced entrepreneurs.

5.1.3   The Academic Practices of Catalyzing Innovation

In this chapter, we consider three practices which, if well executed, could better stimulate and capture the richness of the university's creations, enhance their quality and effectively exchange them with partners. The practices are shown in Table 5.1

Table 5.1 The academic practices of catalyzing innovation

Practice icon	Practice name: description of the practice > and its outcome
MATURING CREATIONS	*Maturing discoveries and creations*: Making progressive discoveries, creations, inventions, market analyses, and proof-of-concept demonstrations > yielding creations with higher technology and market readiness
FACILITATING DIALOGUE	*Facilitating dialog and agreements*: Actively facilitating informal dialog and formal agreements with partners > improving understanding of partner needs and enabling more university creations to be adopted by partners
VENTURING	*University-based entrepreneurial venturing*: Engaging in the real entrepreneurial process within the university, supported by networks of mentors, with access to investors and facilities > producing new ventures and more experienced entrepreneurs

and discussed in the remainder of the chapter. We also present seven cases that help define the practices and serve as examples of their effectiveness.

The impact these practices will have at the university and its faculty is best exemplified by the transition from a reference, the average situation at good universities, to an aspirational situation (Table 5.2).

Our three proposed practices will assist a university in reaching these three aspirational situations. The first practice is about *translating creations* to a mature and market-ready form [3] and is the foundational practice of Catalyzing Innovation. Those creations find their way to products and systems *produced by industry and government institutions* primarily through Facilitating Dialog and Agreements. The university creations are *taken up by start-ups* through University-Based Entrepreneurial Venturing. A related educational practice is Preparing for Innovation, which readies students in entrepreneurship, management, and leadership (Chapter 3). There are two related research practices: Centres of Research, Education, and Innovation, which produce directly implementable solutions; and Undergraduate and Postgraduate Student Researchers, which expose students to new discoveries and creations (Chapter 4).

Table 5.2 Reference and aspirational situations for catalyzing innovation

Reference situation for innovation	Aspirational situation for innovation
Faculty routinely publish their research outputs in scholarly publications, but their uptake and impact on solutions developed by partners is difficult to identify	Faculty review their work with experts, identify discoveries and creations with commercial potential, seek IP protection, mature, and demonstrate them, and are rewarded by seeing the impact of their work [Maturing Creations]
Researchers have not had much opportunity to meet with colleagues from partners, or help them adopt creations	Researcher are frequently in contact with colleagues from partners, try to understand their needs, and work to help them adopt creations, including those covered by IP protection [Facilitating Dialog]
Students and faculty sometimes launch start-ups using their own efforts and resources, but the investment in them is modest, and the success rate is below benchmarks	Students and faculty commonly launch start-ups, learn through structured experiences, work with mentors, and have access to facilities and capital—Resulting in more experienced entrepreneurs with higher success rates [Venturing]

5.2   Maturing Discoveries and Creations

MATURING CREATIONS

5.2.1   *Increasing the Maturity of Discoveries and Creations*

The process of moving ideas from lab to market is inhibited by a gap between the lower level of readiness normally achieved by university researchers and the higher level needed by partners [4]. The goal of this practice is to close this "readiness gap," maturing the creation from both a technical and commercial perspective. This proactive process by the university will lower the barriers associated with adoption by industry or entrepreneurial enterprise [5]. Because Maturing Discoveries and Creations is crucial to the uptake of university research, we consider it the *foundational practice* in catalyzing innovation.

The norm in university research is that faculty make a discovery or develop a creation, write a paper and move on. By contrast, partners expect a fairly mature, robust, and scalable creation that can be translated to production. The resulting

> **Box 5.2 Goals of Maturing Discoveries and Creations**
>
> Universities will more effectively stimulate and capture the richness
> of innovation creations and exchange them with partners,
> **By making progressive discoveries, creations, inventions, market analyses,
> and proof-of-concept demonstrations,**
> *Yielding creations with higher technology and market readiness*

"readiness gap" is often too big to cross, and otherwise valuable creations go without uptake. One cause of this gap traces to incentives in academia and the nature of funding schemes. Another cause is the unwillingness of industry to absorb technologies that have not been tried by others, often waiting for small enterprise to take this risk. The university should do its best to close the gap and industry should match its steps.

This readiness gap is distinct from the other obstacles on the path from ideas to commercial impact. The development of a product first absorbs cash from investors before producing returns—the so-called Valley of Death [6, 7]. Once a company has sold the product to a few early adopters, they must engage the mid-market, requiring them to "Cross the Chasm [8]."

Maturing Discoveries and Creations is a progressive process of: (1) invention and intellectual property protection, (2) market and business analysis, and (3) proof-of-concept demonstration (Fig. 5.1). It is initiated by the interest of faculty members in maturing a discovery or creation, which often involves building a prototype or proof-of-principle demonstration and creating intellectual property (IP) or tangible research products (TRP). The latter may include engineering prototypes and drawings, new organisms, software, circuit chips, and other such products.

The practice then seeks to identify how a more mature version of the creation will impact markets, how it will lead to products and systems, and what business opportunities will be created. The practice also involves building a proof-of-concept demonstration, sometimes called translational research, in which a more realistic prototype is developed, whose design is influenced by the IP status and market analysis. The key is to focus on potential impact the creation will make.

This practice will take the participating faculty members far from their academic comfort zone. Success at Maturing Discoveries and Creations will likely involve a

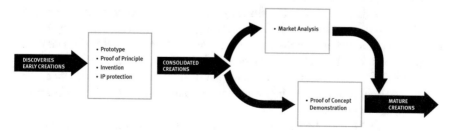

Fig. 5.1 The process of maturing discoveries and creations, involving prototypes, market analysis, and demonstrations

technology savvy business advisor and experts from industry or enterprise, who will advise the faculty and participate in some of the activities.

One way to help creations to cross the readiness gap is to invite industry onto campus to co-create with academic partners. This involves inviting industry partners to take up residence as visiting Joint Research Chairs and even to create their own industrial laboratories within the confines of the university. This is described in the case called Osaka University—Research Alliance Laboratories (Case 5.1).

Case 5.1 Osaka University—Research Alliance Laboratories "Industry on Campus"

Researchers from industry, with appointments to Joint Research Chairs and running Research Alliance Laboratories located on the campus, work with regular university professors and students to develop results of importance to industry.

Osaka University is a comprehensive public university with 16 graduate schools, 11 undergraduate schools, 24,000 students, and 3000 academic staff. It was founded in 1931 based on strong demand from the business and government sectors of Osaka. Today, it is responding to changes in the university environment around globalization, demands in human resources and industry-education collaboration.

One of the campus leaders in industry collaboration is the Graduate School of Engineering, with 6000 students and more than 500 academic staff. The University's program "Industry on Campus" brings about true industry-university collaboration in developing technologies and products, identifying the seeds of new research, and developing human resources.

Prior to 2006, the normal interaction with industry occurred between a researcher in a company and a professor at Osaka University. Since then, two efforts have been launched in the Industry on Campus program. Under a Joint Research Chair, an individual from industry holds a position as a specially appointed professor or associate professor and works in a laboratory on campus with a small team. Based on the success of Joint Research Chairs, an even more extensive program called Research Alliance Laboratories was started in 2011 [1]. Here, an entire section of a company works on campus in a university research building.

Joint Research Chairs are established to conduct joint research coordinated by the university and the sponsor company. The university hosts the researcher on campus, and the company researcher holds a position as a specially appointed professor. This individual is identified by an Osaka University mentor-professor and a company, and the school makes a term appointment. The Osaka University faculty work together with the company researchers on an equal basis. The joint research work of these chairs is characterized by its flexibility and fast pace. As of April 2019, 13 companies were engaged in Joint Research Chairs under the Graduate School of Engineering, including Panasonic, Shimadzu, and Mitsubishi Electric Corporation.

Research Alliance Laboratories are larger scale programs, attracting a company research organization that might include 20–40 industry researchers. Many

Fig. 5.2 Osaka University has taken advantage of multiple joint research initiatives with industry. Research Alliance Laboratories carry out the latest research in a variety of fields

evolve from Joint Research Chairs, and the Osaka University mentor-professor often acts as a codirector. Working together, the company and Osaka University facilitate industrialization of research achievements, improve sophistication of research activities, and nurture human resources. They do this by using mutual research information, technologies, human resources, and facilities at the Research Alliance Laboratory.

By 2019, 11 Research Alliance Laboratories were operating, with significant results. For example, Komatsu is developing the technology for large industrial earth moving equipment, such as teleoperation and autonomous travel of construction machinery. Hitachi Zosen, working through the Hitz(Bio) Research Alliance Laboratory, is developing an eco-functional bio-polymer with superior tensile and impact strength based on plant biomass.

Many of the Joint Research Chairs and Research Alliance laboratories are located in university owned TechnoAlliance building, but some are located elsewhere on campus (see Fig. 5.2). They are usually established for an initial term of two or three years, which can be extended by agreement. Ownership of intellectual property follows normal campus policies; the share of the right is determined based on the contribution of the inventor. Students can also participate and are able to do an "internship on campus," which is industrial internship conducted on campus.

Contributed by Toshihiro Tanaka, Professor, Dean of the Graduate School of Engineering, and Toshitsugu Tanaka, Professor of Mechanical Engineering, Osaka University, Osaka, Japan.

Reference

1. Research Alliance Laboratories. In: Graduate School of Engineering, Osaka University. https://www.eng.osaka-u.ac.jp/en/research/jointresearch.html. Accessed 20 Jan 2020

Box 5.3 Maturing Discoveries and Creations—Practice and Key Actions

Maturing Discoveries and Creations is a progressive approach that matures university discoveries and creations through a process of invention, intellectual property protection, market and business analysis, and proof-of-concept demonstration. These actions raise the technical and market readiness for adoption and increase the likelihood of uptake by industry, enterprise, or government institution. This is the *foundational practice* for Innovation.

Rationale: Moving ideas from the university to partners is inhibited by a gap between the low level of readiness normally achieved at universities and the higher level desired by partners before adoption. This process helps to bridge the "readiness gap" and accelerate the uptake of new ideas. It engages participants from industry and enterprise as market experts and as potential users of the creations.

Key actions:
- Identifying and consolidating a creation with potential market impact by creating a prototype, proof of principle or invention, and seeking IP protection.
- Identifying and assessing the potential commercial impact of the creation in a specific market, product or system application, or business opportunity.
- Conducting a proof-of-concept demonstration to mature the creation and validate it in the context of a potential market, closing the readiness gap.

The key outcomes are more technology-ready and market-ready creations.

Three key actions for implementing this practice are summarized in Box 5.3, and they are each discussed further below.

5.2.2 Consolidating and Maturing Discoveries and Early Creations

New creations arise in different ways. The genesis of a creation is sometimes a research discovery, which sits on the spectrum from curiosity-driven to use-inspired research. Regardless of where a given discovery sits, the faculty researcher can always consider how to impactfully apply the discovery to a societal issue, sometimes leading to a creation (Impactful Fundamental Research in Chapter 4).

Some creations appear spontaneously without any precursor discovery—a new design, piece of code, or even work of literature. With time, the early creations converge to an identifiable and consolidated form. Ideally, each consolidated creation has some foreseeable potential for impact, along with some sort of associated invention, prototype or proof of principle.

Once the creation is identified, the faculty should consider disclosing a piece of intellectual property (IP), such as an invention. Some creations will be covered by a copyright—software, writing, and drawings. Other more tangible items are called Tangible Research Property (TRP). This category includes engineering prototypes and drawings, new organisms, software, circuit chips, etc. Finally, knowledge exchange may involve process knowledge, know-how, concepts, and business ideas. The researchers should understand the criteria, timing, and procedures that must be followed to obtain protection in these various cases, and they should stay in contact with the university office that deals with these issues [9] (Facilitating Dialog and Agreements below).

In order to consolidate the creation, it is important to build an early prototype, incorporating the key elements of the concept and demonstrating functionality. Such proof-of-principle prototypes will help to refine the creation, demonstrate reduction to practice for the purposes of IP filing, and prove valuable in attracting further support and attention from partners and funding sources. These very early prototypes are normally built on the margin by the original team. Sometimes small "ignition grants" are available from the university or government funding agencies.

5.2.3 Identifying Markets for Products Enabled by the Creation

The next two actions are identifying markets and conducting proof-of-concept demonstrations. These run roughly in parallel, but the market effort leads off first. Both draw in cross-disciplinary experts who help to drive the ideas and technologies to innovation solution.

The market exploration effort identifies a commercial product, system, or business opportunity that may be enabled by the creation. This commercial view influences the design of the proof-of-concept demonstration, strengthens IP claims, and informs licensing discussions.

Normally, academics assume that the more difficult problem is increasing the creation's technical readiness level (TRL) [10]. In fact, the commercial assessment can be more difficult [11, 12]. It can expose a gap in business readiness level (BRL). Assessments of creations that improve an existing product are more conventional. Special approaches are needed when a creation leads to a product that meets an unidentified latent need or that involves disruptive technologies.

We cannot expect the faculty to perform this commercial assessment alone. They should team with seasoned market experts and knowledge exchange professionals (Facilitating Dialog and Agreements below). This team seeks a commercial target for the creation. Often the most obvious application is not the best one, and many alternatives need to be considered. It is possible to use this market assessment as a learning experience for students, such as in a class in entrepreneurial marketing (Preparing for Innovation in Chapter 3). Any member of the team might become a founder in a start-up, but this is not an explicit expectation.

The process should go as far as developing a go-to-market strategy, contacting potential commercial partners and engaging them in the next phase: the proof-of-concept demonstration. The more the demonstration is focused on a market application, the more effective it will be in maturing the creation in that direction.

Another successful approach to identify future markets and products is through integrated industrial planning. This is discussed in Case 5.2 titled KAIST—Working with Government and Industry to Mature Technologies. KAIST is able to focus on specific sectors and outcomes which are agreed ahead of time as being in the national economic interest.

Case 5.2 Korea Advanced Institute of Science and Technology (KAIST)— Working with Government and Industry to Mature Technologies

A clear understanding of needs, careful planning and close coordination among the university, industry, and government can contribute to building strong economic sectors.

As a unique university established through a special government law, KAIST focuses on specific fields of research and training that contribute to Korea's development in science and technology, as well as to Korea's industrial growth.

KAIST aims to strengthen the commercial utilization of university research achievements by reducing the readiness gap between idea-based research and the more mature technologies sought by industry. To address this issue, KAIST has implemented a unique and effective process built on the Triple Helix Model.

The three-stage process is ① Needs ② Plan ③ Match (see Fig. 5.3). It involves close cooperation with enterprises and the Korean government and is a KAIST key success factor in translating research to adoption by industry. The three stages are:

① Needs: the identification of new target industries that can become new engines of growth for enterprises and the nation
② Plan: establishing medium and long-term plans for national science and technology policies to grow specific fields
③ Match: providing industries with both R&D and specialized talents that are expected to be in demand among enterprises, by considering national science and technology policies

To see the impact of this three-stage process, consider the fact that KAIST is the main driving force behind the 70% global semiconductor market dominance of Samsung Electronics and SK Hynix.

The Korean semiconductor industry produced R&D achievements through the systematic and strategic collaboration among the government, Samsung Electronics, and KAIST. This became the driving force for growing national wealth for the past 40 years. The three-step model supported this development.

Fig. 5.3 Three stage model for migrating technology to enterprise

The Need: Samsung Electronics decided on semiconductors as its driving force for growth. In the 1970s, Samsung became the first domestic supplier of chips, and in the 1980s it established a semiconductor research center. In the 1980s–1990s, demand for skilled workers and research increased as groups such as Hyundai and LG entered the industry [1]. Today, Samsung leads the world by focusing on system semiconductors.

The Plan: Since the 1970s, semiconductors have been a core driving force of Korea's economy, with the nation achieving a world-leading semiconductor export milestone of 100 billion USD in 2018 (a cumulative total of 1 trillion USD). The Ministry of Trade and Industry has invested and focused on successive waves of this development.

Match: For 40 years, KAIST has been the cradle of semiconductor academic-industry cooperation. In 2018, Samsung Electronics established the KAIST-Samsung Electronics University-Industry Cooperation Center to pioneer the growth of next-generation semiconductors (see Fig. 5.4). KAIST is also an important supplier of talented workers. In some divisions of Samsung, the technical workforce approaches 20% KAIST graduates.

As an institute founded with the goal of leading the growth of the nation's industries through science and technology, KAIST determines the direction of key research fields and talent development programs according to the science and technology policies of the Korean government. The process in which the government establishes policies by taking into consideration the needs of enterprises has continuously and progressively developed in stages across 40 years.

Fig. 5.4 KAIST-Samsung University-Industry Cooperation Center

The cooperation has evolved from one based on demand, then trust, and now, into a new cooperative relationship that produces outstanding achievements.

Contributed by Sung-Chul Shin, President, Korean Advanced Institute of Science and Technology, Daejeon, South Korea.

Reference

1. KAIST Committee (1996) Korea Institute of Science & Technology, Half a Century: A Never-Ending Challenge to the Future

5.2.4 Proof-of-Concept Demonstration

This demonstration seeks to mature the creation so as to reduce the readiness gap, validate it in the context of a potential market, and put on a good show. A proof-of-concept demonstration is probably the last step in maturating a creation within the university. It is built and executed on a timetable that runs parallel to but slightly behind the marketing study, so that its design reflects a potential market application. Before the demonstration, IP ownership should be confirmed, appropriate IP protection should be filed, and agreement reached on who will be the IP licensor.

The audience for the demo is a potential investor in a start-up or a licensee of the IP in industry. They will seek evidence of:

- Functionality, incorporating the creation in a realistic way.
- Scalability to a commercial scale.
- Reproducibility and robustness to variation.

The demonstration should be framed in the context of the identified market and product. It should be convincing and easily explainable. Funds will be necessary for the demonstration, and this will require creative financing by drawing on university funds, government programs, philanthropies, potential investors in start-ups or licensees of the IP.

A social goal of the demonstration is to grow the skills and coherence of the university team. It is desirable that the demonstration be built by the original team, so that the tacit knowledge needed will be available. The skills of the original team can be strengthened by involving others, especially those from different disciplines and professional backgrounds. The effort should involve a full-time or part-time project manager skilled in product development. The demonstration activity will build skills and self-confidence in the team, some of whom may participate in a start-up or in the transfer of the creation to industry.

5.3 Facilitating Dialog and Agreements

5.3.1 The Importance of Knowledge Exchange with Partners

Knowledge Exchange is the multidirectional flow of information at the porous boundary between the university and its partners, in pursuit of shared solutions or to address common needs. It is useful in education and research, but it is critical in the adoption of university creations by partners. This practice outlines the systematic and proactive support that facilitates informal contacts around innovation and manages formal IP processes. This practice is about supporting *adoption* of creations by partners, while the previous one is about *maturing* creations at the university.

Box 5.4 Goals of Facilitating Dialog and Agreements

Universities will more effectively stimulate and capture the richness of innovation creations and exchange them with partners,
By actively facilitating informal dialog and formal agreements with partners,
Improving understanding of partner needs and enabling more university creations to be adopted by partners.

The outcome of Facilitating Dialog is the adoption by partners of university creations. This adoption is aided by long-term relationships of trust and cooperation, which allow faculty and partners to be in frequent contact, and develop mutual understanding and respect. They learn about each other's needs, capabilities, and efforts, and share know-how, leading to informed actions and benefits on both sides. One cycle of success will lead to more efforts. The industrial ecosystem also evolves, building more absorptive capacity to turn creations into marketable innovations. And economic development is likely to be added through partnerships between industry and universities [13].

Understanding the needs of society and partners shows up in all three domains. In education, this discussion normally centers around desired student learning outcomes (Integrated Curriculum in Chapter 3). In research, it helps guide the practice of Impactful Fundamental Research and contributes to the definition of Centres of Research, Education, and Innovation (Chapter 4). In creations, it takes on special significance because of the depth, volume, and level of detailed knowledge being exchanged.

Universities are multifaceted and distributed organizations comprising largely independent scholars. Innovation knowledge exchange is strengthened by three types of contacts with partners: bottom-up, professionally facilitated, and high level. Scholars and colleagues at partners naturally exchange ideas through bottom-up contacts and collegial relationships. This direct contact is essential. This can be complemented by having trained professionals interface with partners, engaging them in discussions around issues that address their needs. The practice can inform industry of recent research discoveries and innovation creations, arrange formal briefings, and help broker deals. It can identify IP of interest and move ahead with formalization of IP and TRP agreements.

At the highest level, major relationships require that university leadership be involved. They deal with their senior counterparts at partners, set strategic agendas, and expedite progress by faculty and knowledge exchange professionals.

A strategic approach to facilitating agreements and dialog is the creation of geographically distributed off-campus platforms, which complement on-campus efforts. An example is presented in the case HKUST—Knowledge Transfer at the Campus and Off-Campus Platforms (Case 5.3). These off-campus platforms facilitate more local knowledge sharing, focused on sectors of interest in the regional economy.

Case 5.3 The Hong Kong University of Science and Technology (HKUST)—Knowledge Transfer at the Campus and Off-Campus Platforms

A university can expand and strengthen its knowledge transfer influence by working from its main campus, and by creating off-campus regional platforms.

The Office of Knowledge Transfer (OKT) and the Fok Ying Tung Graduate School (FYTGS) are the two HKUST organizations responsible for Knowledge Transfer endeavors (see Fig. 5.5). Both are under the Associate Vice-President

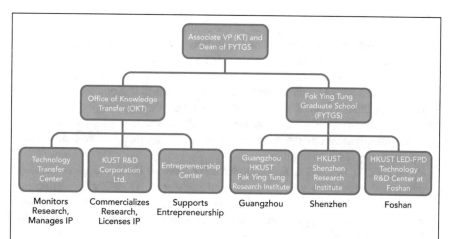

Fig. 5.5 Organization of Hong Kong University of Science and Technology campus knowledge transfer activities and off-campus platforms

(Knowledge Transfer) who is also the FYTGS dean. This arrangement provides overall coordination for knowledge transfer activities at HKUST's Clear Water Bay campus in Hong Kong and Mainland Platforms.

The *Office of Knowledge Transfer* includes the Technology Transfer Center (TTC), the HKUST R and D Corporation Limited (RDCHK), and the Entrepreneurship Center. One of OKT's roles is to work with a range of stakeholders, including government departments, industries, NGOs, faculty, staff, students, and alumni, to promote and facilitate technology transfer endeavors.

The *Technology Transfer Center* serves as a bridge between HKUST research community and the business sector, identifying collaboration opportunities in the local, regional, and international markets. TTC monitors emerging fields of research within HKUST for commercialization potential; manages intellectual properties arising from HKUST research; supports faculty and students in their endeavors to pursue university-industry collaboration, entrepreneurial technology transfer, and commercialization.

HKUST R and D Corporation Limited is a wholly owned HKUST subsidiary and serves as the business arm of the university to commercialize research. Through partnership and knowledge transfer with industry, the company supports the university's role in economic development, mainly in Hong Kong and the neighboring regions.

The *Fok Ying Tung Graduate School*, established in 2007, oversees the university's knowledge transfer endeavors at HKUST's mainland platforms in Shenzhen, Guangzhou, and Foshan. FYTGS's mission is to ensure that the university's efforts to advance innovation and entrepreneurship in Mainland China are integrated and coordinated.

The *Guangzhou HKUST Fok Ying Tung Research Institute* engages in innovative research, product development, training, entrepreneurship, and commercialization to promote the development of the Mainland economy. It focuses on the research and development of internet of things, advanced manufacture and automation, advanced materials, and sustainable development.

The *HKUST Shenzhen Research Institute* supports academic and research collaborations with Mainland universities and institutes. It promotes technology innovation, industry collaboration, and entrepreneurship. Research and development includes biomedicine, advanced materials, new energy, environmental engineering, marine environment, marine biology, automatic control, advanced manufacturing, and smart cities.

The *HKUST LED-FPD Technology R&D Center at Foshan (Foshan Center)* addresses the technology development and transfer of HKUST's light-emitting diode and the flat-panel display to industry partners. The Foshan Center also works in areas including advanced opto-electronics device and component manufacturing and materials.

The performance of this array of organizations at both Clear Water Bay campus and Mainland Platforms has been strong. In academic year 2017–2018, 244 patent applications were filed and 143 granted, yielding a current IP portfolio of 1390. In the same year, the total income generated through applied research, contract research, consultancy and testing services, including those from the Mainland Platforms, reached HKD 285 million [1].

Further advancing HKUST's innovation and interdisciplinary agenda, a new campus, HKUST (GZ), is being established in Guangzhou. Situated in Guangzhou's Nansha district and supported by convenient transportation, the new establishment will enable HKUST to synergize both campuses with new academic degrees and research programs, supported by substantial new resources.

Prepared based on input provided by The Hong Kong University of Science and Technology, Hong Kong, China.

Reference

1. Hong Kong University of Science and Technology (2018) Knowledge Transfer Annual Report 2017/18. https://kt.ust.hk/sites/default/files/reports/file/HKUST17.pdf. Accessed 20 Jan 2020

In all of this, the importance of the human dimension of Knowledge Exchange cannot be overemphasized. Knowledge Exchange is best done by mutually respected counterparts who understand each other, visit or talk often, share experiences, and trust one another. Formal policies (incentives and institutional arrangement, publications and IP agreements) matter. But the real process of Knowledge Exchange is essentially an informal human activity.

Four key actions for implementing this practice are summarized in Box 5.5, and they are each discussed further below.

> **Box 5.5 Facilitating Dialog and Agreements—Practice and Key Actions**
>
> *Facilitating Dialog and Agreements* provides systematic and proactive support of the effective adoption of university creations by partners. It includes technology licensing, support for partners learning about creations, and facilitation of informal multidirectional exchanges that are long-term platforms for informed actions and the sharing of know-how.
>
> *Rationale:* Universities are distributed developers of creations. In order to maximize the potential for innovation, there should be multiple pathways for exposing creations and facilitating knowledge exchange. This requires informal interactions, where mutual awareness and trust form the basis for effective exchange. Formal publications and licensing are important, but are only a part of effective knowledge exchange.
>
> Key actions:
> - Universities learning about the long-term needs of companies through systematic dialog with partners.
> - Partners learning about the outcomes of the university through events and demonstration days, guided visits, survey presentations, and access to digital archives.
> - Exchanging knowledge with partners through open and trustful long-term relationships, growing from discussions, joint projects, and personnel exchange.
> - Exchanging knowledge with partners through the formal mechanism of agreements on IP (Intellectual Property) and TRP (Tangible Research Property) and funding.
>
> The key outcome is the adoption of discoveries and creations by partners.

5.3.2 Systematic Dialog to Understand the Long-Term Needs of Partners

Universities benefit from engaging with their external stakeholders and learning about their needs and plans. This could lead to innovation activities that are more relevant to the needs of society, more likely to have impact on economic development and to attract resources. There needs to be a dialog, bringing together scholars with thought leaders in industry, enterprise and government organization, with sufficient time and facilitation to allow sharing of ideas. This process need not be only bilateral. The university can also use its broad convening power, inviting various partners, economic disrupters, regulators, and funders to build consensus on needs and visions of the future.

A structured process can uncover unserved needs or those that require longer timescales. The objective is not to agree to short-term service level agreements. The real goals are long-term alignment and identification of new areas of cooperation, leading to agenda setting at the university, at partners, and in policy. At such a level, strategic agreements and funding arrangements can be struck.

The important elements of such a structured dialog process include:

(a) The university leadership takes the initiative to identify a key issue, joined by responsive leaders from partners, who then draw in relevant faculty and colleagues from partners.
(b) An agreed agenda is followed by workshops with skilled facilitation, involving task or project groups, which begin to build relationships around knowledge.
(c) Outcomes are summarized in a report by an interested but un-conflicted individual, ensuring that all views are fairly reflected.

Of course the real benefit comes when parties take action on the findings.

There are other systematic ways that universities can learn the needs of partners, including grand challenges set by government and the Sustainable Development Goals of the UN. Industry sometimes sets up online platforms with their needs, and government agencies often bundle problems in announcements.

5.3.3   Partners Learning about Recent University Outcomes

Partners benefit when they learn more about the recent relevant outcomes of the university. Aside from hiring students, most partners come to universities to learn about general trends or investigate some specific question. They want to know what others know but they do not, and to identify significant holes in their fabric of knowledge. The university benefits by providing a multi-tiered approach, first to inform partners and then to engage them.

The basic approach involves some faculty communications targeted at informing partners. Broad partner awareness is raised when faculty present talks at conferences attended by partners, write in publications with broader readership, and teach professional education courses. Many universities hold events and open demonstration days to showcase recent creations. Web publication of recent advances in digital repositories can be instrumental.

The "premium" approach is to provide managed services that facilitates connections between partner participants and scholars. Such a service can:

• Maintain an inventory of recent university outcomes.
• Help industry make contact with faculty who have results important to them.
• Arrange guided visits at the university, setting up tailored itineraries and arranging for formal briefings.
• Host periodic conferences or online events with survey or specialist presentations by thought leaders.
• Provide access to proprietary digital archives and summary overview reports.

This process is facilitated by professionals in Knowledge Exchange, who are important to the long-term success of the university in its mission to support innovation and economic development. This additional effort has a cost, often shared with participating partners.

An example of a successful program that facilitates mutually beneficial dialog is presented in the case called MIT—the Industrial Liaison Program (Case 5.4). Here knowledge transfer professionals provide services tailored to the company, guiding them through MIT, introducing them to relevant scholars and projects, and facilitating deeper and prolonged involvement.

Case 5.4 The Massachusetts Institute of Technology (MIT)—the Industrial Liaison Program

When tailored for industry, professional liaison services strengthen the relationships between a company and the university, benefiting both.

The mission of the MIT Industrial Liaison Program (ILP) is to create and strengthen mutually beneficial relationships between MIT and corporations worldwide [1]. The ILP is one of several key elements in the Institute's multifaceted and collaborative network of activities for engaging the private sector. The ILP offers its corporate members a professional and differentiated set of products and services that enhance companies' ability to access the resources of MIT.

MIT is a highly complex and decentralized institution where new labs, initiatives, consortia, and centers regularly appear. The ILP's staff of more than 30 industry liaisons provides strategic, ongoing, facilitated access to MIT's 3500 faculty and research staff. ILP industry liaisons have cultivated strong working relationships with MIT faculty to understand their latest research, and perhaps more importantly, their future research interests. Together, this MIT-industry team provides a unified, current knowledge of MIT unlike any other. In addition, the ILP provides targeted access for its members to more than 1700 MIT-connected start-ups through the MIT Start-up Exchange.

The ILP organizes customized executive briefings at MIT for member companies. These are often used to enhance corporate strategic planning, to examine emerging research and technology, and to consider new management approaches to corporate issues. The ILP regularly organizes a set of conferences and seminars where the MIT community and ILP members may share knowledge and network with each other. These events are content-heavy, featuring thought leaders from industry and MIT. In the last year, the conferences have spanned digital health, materials, consumer dynamics, and information and communications technology.

In addition, ILP facilitates other interactions. Industry liaisons have familiarity with MIT systems and processes (research contracts, licensing options, event planning, trademarks, administrative policies, etc.). For those companies interested in recruiting MIT students, ILP industry liaisons are experts at enabling a strategic recruiting process. The ILP also tracks and leverages the surrounding Boston area innovation ecosystem for its members.

Table 5.3 Tiers of corporate relationships

Institute Initiatives
Partnership Companies
Major Programs and Consortia
Portfolio Investors
Focused Research Projects
Consortia and Center Memberships
Access, Meetings, Conferences
ILP Membership

This set of products and services defines the first tier of potentially deepening relationships between MIT and corporations (see Table 5.3). Many companies, having joined the ILP, go on to sponsor focused research programs, participate in major programs, and become partners in major MIT initiatives. ILP's approach to enhancing MIT's corporate relations is a holistic, relationship-driven model.

At the same time, ILP provides valuable service to MIT faculty leaders. The ILP liaisons can arrange introductory meeting with potential industry-funded research sponsors, so that faculty more directly learn the needs of industry. ILP can also help arrange translation pathways suggesting how previous research can gain impact in industry. Faculty support is vital to ILP's success. These efforts are constantly monitored through faculty surveys, in order to measure results and identify improvement opportunities.

ILP staff also support members of MIT's senior administration in their corporate relations activities, including the pursuit of major MIT international and corporate partnerships. By having a critical mass of internal resources, the ILP actively supports faculty and institutional initiatives.

Contributed by Karl Koster, Executive Director, MIT Corporate Relations, Massachusetts Institute of Technology, Cambridge, Massachusetts, USA.

Reference

1. MIT Corporate Relations. https://web.mit.edu/industry/ocr.html. Accessed 20 Jan 2020

5.3.4   Exchanging Creations with Partners through Open Informal Relationships

When industries learn about the relevant outcomes of a university, they develop more interest and are drawn into closer interaction. Building on cycles of introductions, discussions, and then coordinated efforts, these activities lead to open and trustful relationships based on mutual knowledge and respect. We stress the importance of these recurring informal human interactions as the basis of effective knowledge exchange.

Alumni and other professional networks often develop near the university or around a discovery or creation. Such informal professional social networks are important in disseminating materials, especially in the entrepreneurial community (University-Based Entrepreneurial Venturing below). The university's convening power and transfer of knowledge through its students and informal networks are important in stimulating local economies [14].

Growing personal contacts provide an opportunity for colleagues in the university and partners to exchange know-how and tangible research products. It also establishes contact between the partner and research students, who may eventually become employees. Sometimes this contact leads to consulting relationships at the partner by the faculty as a way to support adoption of a technology.

The informal professional friendships and networks are augmented by approaches with a more organizational basis:

- Exchange of academic and industrial staff.
- Creation of academic positions for distinguished practitioners from industry; these are sometimes called professors of practice.
- Colocation of projects, often in university labs, and sometimes involving partner participants.
- Consortia of interested parties, often including partners, that can coordinate funding, planning, and execution.
- Corporate laboratories near the university, facilitating interaction, sometimes with part-time faculty appointments.

This variety is needed because different partners respond to different models of collaboration.

5.3.5 Exchanging Creations with Industry through Formal Agreements

An important aspect of the university's knowledge exchange program is formal agreement on intellectual property (IP-like inventions and copyrights) and tangible research property (TRP such as engineering prototypes and drawings, new organisms, software, and circuit chips). The governing principle is that IP should be made available to partners and others in the public interest, while respecting the rights of the inventors. The pursuit of formal IP protection should not stand in the way of prompt and open dissemination of research and innovation outcomes.

These relationships are often managed by knowledge exchange professionals at a Technology Transfer Office (TTO). These professionals scan for potential IP among research groups, secure IP protection, and reach licensing agreements. Sometimes these agreements are with university start-ups. Such start-ups further mature the technologies and themselves become instruments of knowledge exchange in the commercial ecosystem.

Reporting metrics often emphasize these formal knowledge exchange mechanisms. Although a focus on the number of agreements or short-term income generation is tempting, it can be counterproductive. It can complicate important partner relationships and cause the university to miss the point that IP should primarily be an instrument of effective knowledge exchange and not of operational funding.

Each university, influenced by national policy and law, should develop its strategy to facilitate this form of knowledge exchange. Depending on the circumstances, IP may be owned by the university, professor, or sponsor. It is important that IP ownership is clear and that the owners have agreed on who will be the IP licensor. The TTO's decisions should be forward looking, as intellectual property may be produced well before industry is able to adopt it.

Another form of formal agreement between partners and the university is corporate funding for research and innovation. In some nations, funding from foundations and partners forms a substantial fraction of the total funds available to the university. Corporate funding, with active partner involvement, is a strong incentive for informal open interaction, and signals the important issues of societal and partner interest.

Case 5.5 titled Trinity College Dublin—Knowledge Transfer: Research to Impact exemplifies many key points. TCD has a well-organized process of licensing of new technologies and developing collaborative research programs with industry. But it emphasizes the social and economic impact created by such efforts, rather than more traditional metrics of short-term financial return to the university.

Case 5.5 Trinity College Dublin (TCD)—Knowledge Transfer: Research to Impact

Trinity College Dublin emphasizes the translation of research findings to enable economic and social impact, rather than more conventional measures of revenue and economic return.

Trinity College Dublin is dedicated to the principle that research, pulled through to innovation and impact, is better managed for the benefit of society than solely for financial return to the university. The economic aspect of this influence can be achieved through many mechanisms, including the creation and incubation of new businesses and the training of graduates to drive innovation. But in this discussion, we will focus on support of existing companies through the licensing of new technologies and the development of collaborative research programs, increasing company competitiveness and feeding their innovation pipelines.

Trinity Research and Innovation (TRI) interfaces internally with Trinity College academics and administration units, and externally with industry, funding agencies and government bodies. It provides support and advice along the continuum from research funding application, to contract negotiation, intellectual property management, and through to licensing. Trinity has successfully connected its research excellence to industry and has engagements with more than 450 companies.

TRI comprises three offices. The Office of Corporate Partnership and Knowledge Exchange (OCPKE) supports both industry engagement and the

How to work with Trinity
Methods of Engagement

Enterprise Ireland	Science Foundation Ireland	Direct Industry Support	European Union	IDA	Higher Education Authority
Technology Centres	SFI Investigators Programme	Contract Research	Framework Programme	Securing FDI	Delivery of Infrastructure Access
Innovation Partnerships	SFI Industry Fellowships	Rapid Materials Analysis	Partnering with Leading Institutions	Promoting Irish Research Excellence	Enabling Fundamental Research Base
Innovation Vouchers	SFI Research Centers	Local Manufacturing Support	Leveraging EU Funding	Showcase the "Best in Class" Technologies	Foundation Delivery of Core Research
Technical Feasibility Study	SFI Research Centre Spokes Programme	Consultancy	Networking Opportunities	International Company Engagement	
Commercial-ization Fund	Tech Innovation Development Award (TIDA)	Specialist Training	Service Agreements	Technology Centre	
	SFI Strategic Partnerships		Horizon 2020	Training Support	
	SFI Short-Term Industry Visiting Fellowships		Recruitment, Placement & Exchange		

Information is subject to change

Fig. 5.6 Trinity Research and Innovation helps companies navigate the complicated funding system to identify the right opportunity for the proposed partnership

commercialization of Trinity research. It has a simple operating principle—creating a single decision-making entity for all interactions with industry. It provides a pathway between faculty and enterprise that is simple, flexible, professional, and proactive. The Research Development Office maximizes the competitive research funding into the College while supporting the strategic development and implementation of research policy and systems. The Contracts Office has responsibility for reviewing and executing all funded research contracts (see Fig. 5.6).

Trinity College Dublin measures itself by how successful it is in translating research to achieve impact. It simplifies industry partnership by providing a single open door for companies to engage in collaborative research programs and access highly trained scientists, state-of-the-art labs, and dedicated funding specialists. The process starts by listening and dialog: what is the company's need and its problem statement? It progresses through the identification of a research lead, a piece of intellectual property, or a research competency that enables a solution. The single organizational unit ensures a fast turnaround on projects. It provides transparency in identifying barriers to collaboration and a simple escalation path to remove them. Trinity is committed to partnering with and servicing industry through flexible solutions, from training to entrepreneurial programs.

The programs developed by Trinity have inspired the Irish community to focus more on linking knowledge transfer and entrepreneurship. For example, the model of the OCPKE, with a clear focus on enterprise partnership and spin-out creation, has resulted in other Irish universities changing their approach to these practices.

In addition, Trinity is spearheading the development of the Grand Canal Innovation District, in collaboration with other Dublin universities and the broader enterprise community. The innovation district will be centered on a new innovation campus that will bring together at scale university research, start-up and scale-up companies, and large innovative businesses. Collaborations will be enhanced by high innovation density and critical mass. The innovation district will provide an international focal point for Ireland's innovation strategy and unlock the full innovation potential of Dublin. It will be a vital step in positioning Dublin to be ranked as a top 20 global city for innovation.

Contributed by Diarmuid O'Brien, Chief Innovation and Enterprise Officer, Trinity College, Dublin, Ireland.

5.4   University-Based Entrepreneurial Venturing

5.4.1   Becoming an Entrepreneur

An important factor in the economic impact of universities is the contribution by entrepreneurial businesses created by faculty, staff, students, post docs, and alumni. The impact of these businesses can stretch over decades, creating products that benefit society, and providing rewarding employment. Here, we focus explicitly on the preparation of new entrepreneurs—faculty, staff, students, and post docs—and the early companies they will start.

> ### Box 5.6   Goals of University-Based Entrepreneurial Venturing
>
> Universities will more effectively stimulate and capture the richness of innovation creations and exchange them with partners,
> **By engaging in the real entrepreneurial process within the university, supported by networks of mentors, with access to investors and facilities,**
> *Producing new ventures and more experienced entrepreneurs.*

Start-up ventures are a legitimate part of the university's innovation mission. Their success will be valued by stakeholders, justifying significant investments in research and catalyzing innovation. The university will especially benefit if it takes a long view of the relationship with a start-up. Rather than emphasizing shorter term revenue generation, the university should seek good relationships that may lead to reinvestment in the university.

It is challenging to be a successful new entrepreneur. The university can create programs and incentives that stimulate entrepreneurial interest and build entrepreneurial knowledge and skills, developing both capability and self-efficacy. Observers can identify patterns of action that increase the likelihood of successful entrepreneurship, and those that may lead to common mistakes [15]. New entrepreneurs need to learn the successful patterns [16]. It is desirable to have a framework that codifies these successful patterns and does not rely on participants piecing together the principles by themselves from unrelated and sometimes conflicting anecdotes told by successful entrepreneurs.

These patterns can be learned by engaging the learners in the real venture creation process, with mentoring by experienced entrepreneurs (Fig. 5.7). The university provides facilitated access to IP and know-how. Faculty bring deep knowledge of discoveries and creations. Students contribute a breadth of contemporary knowledge, skills, and boundless enthusiasm. Post docs are especially important because of their maturity and the ability to focus on an innovation project. They are more likely to have experience in leadership and management of technology. Universities should consider creating a rank of *innovation post doc* alongside the conventional research post doc.

Universities can facilitate an ecosystem of support for entrepreneurship, together with access to professional networks—what the MIT Martin Trust Center calls "expertise, support and connections [17, 18]." The university should try to build a fertile environment for the development of entrepreneurship. The university can facilitate tangible support in the form of incubators and capital facilities [19]. Professional entrepreneurial networks can act as platforms for sharing insights, ideas, and resources and can lead to needed support services and funding.

Four key actions for implementing this practice are summarized in Box 5.7, and they are each discussed further below.

Box 5.7 University-Based Entrepreneurial Venturing—Practice and Key Actions

University-Based Entrepreneurial Venturing supports the process of venture creation within the university by faculty, staff, students, and post docs, through development and mentoring of new entrepreneurs, and through university facilitated access to discoveries and creation, incubators and facilities, seed funding and professional entrepreneurial networks.

Rationale: Entrepreneurship is an important mechanism of economic development. Comprehensive support by the university can stimulate greater interest by participants, provide them an understanding of their entrepreneurial strengths and aptitudes, and build entrepreneurial competence and self-efficacy. It can yield additional pathways, from ideas to impact and more successful new ventures.

Key actions:
- Developing new entrepreneurs through a combination of structured frameworks, repeated pre-entrepreneurial experiences, and mentoring.
- Enabling successful new ventures based on knowledge outputs, the expertise of the faculty, and the integrative and enthusiastic efforts of students.
- Providing access to incubators, workshops, capital equipment, and small amounts of university or investor funding.
- Supporting professional entrepreneurial networks designed to share insights among entrepreneurs, faculty, staff, students, post docs, alumni, investors, and suppliers.

The key outcomes are the development of creations and new ventures, as well as potential new entrepreneurs.

5.4.2   Frameworks, Experiences, and Mentoring in Entrepreneurship

Becoming a successful entrepreneur requires experience. A university can contribute to the professional development of its faculty, staff, students, post docs, and alumni by providing a framework of practice, a series of team pre-entrepreneurial experiences, and mentorship.

There can be moments of sheer technical brilliance and stunning market insight in the life of an entrepreneur, but much of the day is routine. Inexperienced entrepreneurs routinely fail in this routine. They would be more successful if they learned a structured framework of the basics; for example, identifying customers, prototyping, and evaluating business opportunity. The alternative to a structured process is anecdotes by successful entrepreneurs. But the plural of anecdotes is not evidence. Hearing successful entrepreneurs describe their successes and failures brings richness to a framework, but, alone, it does not provide one.

Fig. 5.7 The process of maturing discoveries and creations, involving prototypes, market analysis, and demonstrations

Students are likely to cross-pollinate ideas in business plan competitions and other types of student activities which emulate the start-up process. Students often participate in student-initiated projects, hackathons, and other student activities that explore how to mature creation and apply them to solving problems.

Experience matters in start-ups. Repeated previous experience is important, particularly in enterprise processes. In the university setting, participants can go through the steps of forming a company, with little cost of failure. If one or two cycles are done quickly in the university, the next ones are more likely to succeed. The entrepreneurs will learn the value of failing fast [20, 21]. These experiences often lead to involvement in more professional accelerator programs.

Working in a team demonstrably increases the likelihood of entrepreneurial success [22, 23]. Swapping in and out of various teams will inform participants of the roles they can play and when they should find partners. Repeated team efforts some-times also identify the nucleating team of the next new venture.

There is no replacement for experienced mentors advising a new entrepreneur. They provide a sounding board and a source of alternative approaches to an issue. They are often the gateway to the local networks. Faculty and alumni who have been entrepreneurs and members of the entrepreneurial ecosystem are often willing to volunteer as mentors.

5.4.3   Building Successful Ventures Based on Discoveries and Creations

The emphasis of action now shifts to creation of actual legally established ventures, conducted within or closely alongside the university. This is a valid activity of the university if the emphasis is on using the new venture as a carrier of knowledge exchange and not just as a source of financial return. The university assets that can catalyze a start-up are: a deep understanding of, and access to creations and discov-eries; IP, tangible research products, and know-how; accompanied by a potent com-bination of students and faculty.

University-based entrepreneurial projects often begin with a unique understand-ing of some discovery or creation, perhaps combined with an idea that will improve the world, or a passion to have an important impact [24]. Access to discoveries and creations can be arranged through the practice of Facilitating Dialog and Agreements. Some new ventures respond to a pull from the market or society, and some are driven by a creation. In both cases, understanding of technology and market is essential.

Faculty can contribute a deep understanding of the outcomes of their research group. They are experts in their domain, and are often able to foresee its evolution, but are more likely to stay close to their intellectual home. They better understand the inventory of discoveries and creations and the formal IP and TRP of the university. If they have participated in the practice of Maturing Discoveries and Creations, they probably already have a well-developed technology-ready and market-ready idea.

In contrast, students and post docs can contribute based on their broad exposure to knowledge outcomes, gained through the practices of Education in Emerging Thought and Undergraduate and Postgraduate Student Researchers. Students are more likely to experiment cross internal barriers and take more intellectual risk. They will know about their own work, and about the research work of others in their research group and the extended community.

The combination of faculty experts and creative students and post docs is power-ful, but there can be conflicts of interest between:

- The IP rights of the start-up founders and the larger set of inventors.
- A supervising faculty and a student who might both be involved in the start-up.
- The interest of the faculty and university as licensors, and the same faculty as a licensee.

University policy should address these issues.

Taking on these issues, the Technion has developed the DRIVE accelerator, described in the accompanying Case 5.6. The main goal of the DRIVE is transfer technology to the community which might otherwise not leave the lab, and, in doing so, to enlarge the number of start-ups launched by Technion faculty, staff, and students. The target is technologies that are ready, but which lack a concrete commercial and marketing plan.

Case 5.6 Technion—Israel Institute of Technology—The Technion DRIVE Accelerator

Efforts by universities, and in some cases their subsidiaries and partners, to support and finance the entrepreneurial process can have significant influence on the success of the ventures and on local economic development.

The Technion, Israel Institute of Technology, was Israel's first university. Graduates make up the majority of Israeli-educated scientists and engineers, constituting more than 70% of the country's founders and managers of high-tech industries. Technion graduates founded and/or lead two-thirds of Israeli companies on NASDAQ.

The main goal of the DRIVE Accelerator is to increase the number of companies launched by Technion faculty and students, with an emphasis on situations where the technology is ready, but assistance is required to prepare the start-up to attract investors, and to finalize the full commercialization plans [1]. The DRIVE is focused on promoting the transfer of technology, which might otherwise never leave the laboratory, to the community.

The DRIVE Accelerator assists entrepreneurs in the early (pre-seed and seed) stages life cycle of their start-ups. In addition to joining and benefiting from Technion's unique ecosystem, innovative capabilities and strong human capital (researchers, students, and alumni), start-ups may access the Technion's research facilities, infrastructure, and equipment. Access to Technion labs and research centers is provided using the same approach as for established industry, which often conducts funded research at the Institute.

Companies accepted to the DRIVE are offered office space and are eligible for an investment of up to 100,000 USD registered as a convertible loan or SAFE (Simple Agreement for Future Equity). Technion and the Technion Research & Development Foundation (TRDF) receive preemptive rights to participate in future investment rounds. Entrepreneurs learn from world-class mentors who have had significant roles in successful companies and who provide long-term guidance on vision and strategy. Mentors work pro bono, and the DRIVE coordinates the mentor network.

The DRIVE is a part of the Technion's Technology Transfer Office, a TRDF division. In contrast to similar efforts operating within the university, TRDF is a for-profit entity, which is owned by the Technion. The DRIVE's deal flow comes from entrepreneurs who commercialize Technion IP and from alumni. Technion offers its inventors complete commercialization packages that include IP protection and commercialization strategies.

An experienced technology entrepreneur runs the DRIVE. A yearly budget allows some co-investment funding with partners who participate in the investment committee, which is composed of Technion representatives, representatives from industry and partner investors.

The program started in March of 2017. The key indicator of success is whether a company raised additional funding before the end of support from the DRIVE program. The start-ups in the portfolio have received on average $1 million further investment and some received over $6 million.

Prepared with input from Shuli Shwartz, Rona Samler, and Wayne D. Kaplan; Technion—Israel Institute of Technology, Haifa, Israel.

Reference

1. Technion DRIVE Accelerator. https://www.techniondrive.com. Accessed 20 Jan 2020

5.4.4 Incubators, Facilities, and Seed Funding for Start-ups

On day one of a start-up, there is nothing—it is all done from zero. Getting started can be daunting, particularly for those that are creating a start-up for the first time. The university can help by providing access to incubator space, workshops, shared capital equipment, and potentially seed funding. These do not have to be directly provided by the university, but can involve not-for-profit, for-profit, and government resources as well.

Many initial needs can be met on a shared basis by a well-run incubator. At first, the entrepreneur needs table space and IT support, and then the legal, tax, and banking relationships. Soon there will be a need for hiring support and financial accounting, where time-sharing arrangements are essential for small companies.

For anything but a software company, there will be need for workshop facilities that support mechanical, electronic, chemical, and biological development. The start-up can create and operate such facilities, but this is unlikely until a robust round of financing. In the interim, access to shared workshops and associated capital facilities is a great benefit, especially if it includes access to experts on the equipment.

The real challenge comes when the start-up needs specialized equipment with large capital costs; for example, materials handling or wet labs. Recently, some universities have recognized that if they are to create anything but IT start-ups, they need shared major capital facilities. In contrast with high clockspeed sectors (IT and consumer products), more capital-intensive creations in slower clockspeed sectors (e.g., biotechnology, energy, aerospace) will have a more complex path to market, and therefore will benefit from this kind of support.

Eventually, the start-up will need investors, who will at first invest small amounts of money—pre-seed or seed rounds. These early investment funds can come from investment programs of the university or from investors closely affiliated with the university and the local ecosystem [25]. In some countries, government and foundation co-investment schemes help as well.

Incubators can be a force for innovation and for community development, as discussed in Case 5.7 called WITS—the Tshimologong Digital Innovation Precinct. This project locates a digital technology hub in the urban core, where people learn, work, and innovate. Engaging both the university community and those in the core, the dual goals of the project are to create new companies based on digital technology and to develop a skilled IT workforce.

Case 5.7   The University of the Witwatersrand in Johannesburg (WITS)—the Tshimologong Digital Innovation Precinct

The project involves reimagining digital innovation for a future more equitable society, creating opportunities for entrepreneurship both for the university community and for those in the urban core, and developing a skilled workforce.

The concept of "Digital Technology Hubs" will be familiar to many in the USA and Europe. Yet by 2014, Johannesburg, Africa's most important economic center, still did not have such an area where people could gather to learn, work, and innovate. The idea grew into a zone that links hardware, software, and digital content innovation with commercial success; a center where technology businesses and digital products could be nurtured, and ideas foment viable businesses.

In 2018, the Tshimologong Precinct became fully operational as a digital innovation ecosystem in Braamfontein, a central Johannesburg neighborhood, through collaboration among the University of the Witwatersrand's Joburg Centre for Software Engineering, the Province of Gauteng, and the business sector including IBM, Cisco, Telkom, Microsoft, and MMI, among others. The considerations for designing the hub were proximity to a major research university, existing urban infrastructure, the likelihood of government support for scale-up, and the potential to develop an ecosystem fostering business and desirable life style.

Tshimologong, the Setswana word for "place of new beginnings," is an urban community of progressive thinkers reimagining digital innovation for a future more equitable society. Curiosity and complex interdisciplinary challenges drive members to imagine new possibilities. Through research and collaboration, people play, make, learn, incubate, and innovate, while promoting the regeneration of Johannesburg's inner-city area. Direct guidance and advice from experienced technical and business mentors, alongside the robust, effective, and hands-on-experience of the precinct's structured programs—ideation, incubation, and acceleration—support ongoing learning and growth (see Table 5.4).

Table 5.4 Precinct's structured programs

Program	Description
Ideation	Entrepreneurs assess the technical viability of the suggested solution, while identifying the existence of a market for the product or service being developed
Incubation	Entrepreneurs design a business model and approach the market place systematically to test the validity of their most basic assumptions using the Lean Start-up Methodology and Design Thinking
Acceleration	Entrepreneurs develop enhanced business models, products, and services to further grow their businesses in order to exponentially scale post-revenue by connecting with funders, global markets, and high-impact networks

Fig. 5.8 Digital Skills Academy model

The aim was to benefit not only the University's own students, but to become the focal point for digital innovation for all of Johannesburg's citizens: young and old, employed and unemployed, South Africans, and other Africans. Men and women from humble backgrounds that have an idea will get a place to implement the concept and perhaps change the world.

The Digital Skills Pipeline program provides free training to unemployed young people from disadvantaged backgrounds. Some of these programs are designed to expose women and girls to opportunities in the ICT sector, drawing on existing partnerships. These programs offer basic coding skills covering an array of programing languages, including JAVA and Microsoft C#; networking; cyber security; and digital animation. These courses are part of the integrated training and development process that dominates the Digital Skills Academy (see Fig. 5.8).

In just a few months, the community created more than 60 start-ups, with up to four participants each, at various stages of their evolution. The skills development programs bring approximately 150 students into the precinct on a daily basis. Ad hoc activities bring, on average, 50–100 additional visitors each day. The center will contribute to fulfill "Africa Rising" destiny by producing world-class developers, entrepreneurs, and innovators of digital technology, while strengthening University-based incubators across the continent.

Contributed by Barry Dwolatzky, Professor Emeritus in the School of Electrical and Information Engineering, University of the Witwatersrand, Johannesburg, South Africa.

5.4.5 Professional Entrepreneurial Networks

Entrepreneurs naturally form professional networks. These networks are often long enduring and are key to the success of local ventures. They are important contributors to informal knowledge exchange and allow the sharing of insights, business intelligence, new discoveries, and creations. Networks are particularly important for entrepreneurs who do not have access to the large organizations, infrastructure, and historical relations that more established firms draw upon.

In addition to entrepreneurs and innovators, network participants can include faculty, staff, students, post docs, and alumni. Participation of local venture investors is common, as well as by people from the legal, financial, and human resource professions. The network can be self-assembled, but some have a convener, often the operator of the incubator. At some universities, alumni in other cities form university affiliated networks in their city. The university can back these networks by sponsoring the convener, providing logistical support and encouraging the faculty to participate.

It is important that these groups are open and available to professionals, with a low cost of participation. They enable entrepreneurs to leverage network effects, so that the value of the network increases as it grows and meetings become more productive. Entrepreneurs should find that they are developing important professional relationships, some of which lead to advice, resources, and deals. The networks are made up of professionals with mutual self-interest.

Some programmed events can add value to the network if they provide content that is responsive to their evolving needs and is at the state-of-the-art of practice. Entrepreneurs prefer informal, active meetings and real-time discussions rather than traditional conferences. The events stimulate discussion focused on peer-level sharing of best practices, skills, and knowledge of resources.

5.5   Chapter Summary

In this chapter, we present three practices that contribute to economic development by accelerating innovation and entrepreneurship (Fig. 5.9). The practices produce creations and stimulate effective exchange with partners in industry, enterprise, and government institutions. These practices support the principle of systematic knowledge exchange of Box 2.3.

Maturing Discoveries and Creations is a practice that matures research outputs through prototyping, IP protection, market analysis, and proof-of-concept demonstration. The inputs are the discoveries and less mature creations developed under the research practices. The outcomes are more technology-ready and market-ready creation. If this practice leads to transfer of IP and know-how to an established firm or government organization, this is called Facilitating Dialog and Agreements. If this practice leads to a start-up, then the action shifts to University-Based Entrepreneurial Venturing.

Facilitating Dialog and Agreements builds relationships with partners and supports knowledge exchange. It includes scholars learning the needs of partners, partners learning about university discoveries and creations, technology licensing, and informal multidirectional dialog. Inputs to this practice come primarily from the research domain, and particularly from the practice of Maturing Discoveries and Creations. The outcome is the successful adoption by industry and entrepreneurial ventures of the university's discoveries and creations.

University-Based Entrepreneurial Venturing plays the dual role of advancing the development of new university-based entrepreneurs and spinning off entrepreneurial ventures. It takes as inputs the human capital of students who have participated in the educational practices, and the discoveries and creations of the research practices. It is aided by Maturing Discoveries and Creations and Facilitating Dialog and

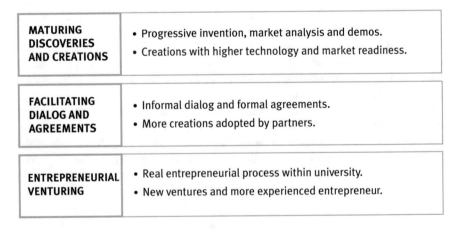

Fig. 5.9 Catalyzing innovation practices, processes, and outcomes

Agreements. In other words, virtually all of the practices of a university with a mission to accelerate innovation influence University-Based Entrepreneurial Venturing.

These three practices are similar to the conclusions of an MIT study that investigated Best Practices for Industry University Collaboration [26]. This study identified seven factors that increase partner adoption of university discoveries and creation. The four factors relevant to interactions between partner and university are shown in Box 5.8.

Box 5.8 Factors Leading to Partner Adoption of University Discoveries and Creations

These factors have been identified as leading to adoption and impact of university creations and discoveries at partners:

(a) Sharing with the university research team the vision of how the collaboration can help the company. This gives the university researcher an understanding of the business context, company practices, and how the research fits in company strategy.

(b) Investing in long-term relationships. The creation of multi-year research plans helps soften the impact of the markedly different timescales of development at the university and in industry. Over time, both sides develop common understanding and vocabulary, and success in one endeavor is highly correlated with success in the next.

(c) Establishing strong two-way communication linkages. Frequent discussions and exchange of personnel in both directions helps form personal relationships and fosters exchange of tacit knowledge and know-how.

(d) Building broad awareness of the project within the company and long-term support until the creation can be exploited. A main point of contact or manager in the company is vital, but broader contact is also beneficial, allowing more widespread sharing of information.

Sustainable economic development can be an important result of the application of the three innovation practices. The practice of Maturing Discoveries and Creations will engage scholars in increasing the readiness of research and innovation outcomes in directions that will encourage economic, social, creative, and environmental good. Facilitating Dialog and University-Based Venturing will increase the likelihood that these university outcomes will impact the market and society.

References

1. Cervantes M. Academic patenting: how universities and public research organizations are using their intellectual property to boost research and spur innovative start-ups. In: World Intelectual Property Organization. https://www.wipo.int/sme/en/documents/academic_patenting.html. Accessed 20 Jan 2020
2. Frølund L, Murray F, Riedel M (2018) Developing successful strategic partnerships with universities. MIT Sloan Manag Rev 59:70–79

3. MIT Deshpande Center for Technological Innovation Evaluating Market Opportunities. http://deshpande.mit.edu/grants-resources/evaluating-market-opportunities. Accessed 11 Oct 2019
4. The Organisation for Economic Co-operation and Development (2013) Commercialising public research: new trends and strategies. Organisation for Economic Co-operation and Development, Paris
5. O'Shea RP, Allen T, Chevalier A, Roche F (2005) Entrepreneurial orientation, technology transfer and spinoff performance of U.S. universities. Res Policy 34:994–1009
6. Markham SK (2002) Moving technologies from lab to market. Res Technol Manage 45:31–42
7. Markham SK, Ward SJ, Aiman-Smith L, Kingon AI (2010) The valley of death as context for role theory in product innovation. J Prod Innov Manag 27:402–417
8. Moore GA (1991) Crossing the chasm. HarperCollins
9. Massachusetts Institute of Technology Guide to The Ownership, Distribution, and Commercial Development of MIT Technology. In: Technology Licensing Office. http://web.mit.edu/tlo/documents/MIT-TLO-ownership-guide.pdf. Accessed 20 Jan 2020
10. Henderson R, Jaffe A, Trajtenberg M (1998) Universities as a source of commercial technology: a detailed analysis of university patenting, 1965–1988. Rev Econ Stat 80:119–127
11. Agrawal A, Henderson R (2002) Putting patents in context: exploring knowledge transfer from MIT. Manag Sci 48:44–60
12. Byers TH, Dorf RC, Nelson A (2015) Technology ventures: from idea to Enterprise. McGraw-Hill, New York, NY
13. Kneller R, Mongeon M, Cope J, Garner C, Ternouth P (2014) Industry-university collaborations in Canada, Japan, the UK and USA—with emphasis on publication freedom and managing the intellectual property lock-up problem. PLoS One 9:e90302
14. Lester R (2005) Universities, innovation, and the competitiveness of local economies: local innovation systems project. MIT, Cambridge, MA
15. Aulet B (2013) Disciplined entrepreneurship: 24 steps to a successful Startup. Wiley, Hoboken, NJ
16. O'Shea RP, Chugh H, Allen T (2008) Determinants and consequences of university spinoff activity: a conceptual framework. J Technol Transf 33:653–666
17. Martin Trust Center for MIT Entrpreneurship (2019) Provides the expertise, support, and connections MIT students need to become effective entrepreneurs. In: Martin Trust Center MIT. https://innovation.mit.edu/resource/martin-trust-center/. Accessed 3 Dec 2019
18. Martin Trust Center for MIT Entrpreneurship (2019) Scaling for impact: pushing the frontiers of entrepreneurship. Martin Trust Center for Entrepreneurship, Cambridge, MA
19. Di Gregorio D, Shane S (2003) Why do some universities generate more start-ups than others? Res Policy 32:209–227
20. Shepherd DA (2004) Educating entrepreneurship students about emotion and learning from failure. Acad Manag Learn Edu 3:274–287
21. McGrath RG (1999) Falling Forward: real options reasoning and entrepreneurial failure. Acad Manag Rev 24:13–30
22. Roberts EB (2018) Celebrating entrepreneurship: a half-century of MIT's growth and impact. MIT School of Management, Cambridge, MA
23. Roberts EB (1991) Entrepreneurs in high technology: Lessons from MIT and beyond. Oxford University Press, New York, NY
24. Aulet B (2017) Disciplined entrepreneurship workbook. Wiley, Hoboken, NJ
25. Etzkowitz H, Klofsten M (2005) The innovating region: toward a theory of knowledge-based regional development. R&D Manag 35:243–255
26. Pertuze J, Calder E, Greitzer E, Lucas W (2010) Best practices for industry-university collaboration. MIT Sloan Manag Rev 51:83–90

Chapter 6
Integrated Knowledge Exchange

6.1 Introduction

The academic practices of a university occur in the three broad domains of education, research, and catalyzing innovation. However, it is the interplay of the practices, the overlap of the domains, and the convergence of disciplines that create many of the unique outcomes important for knowledge exchange with partners. Chapters 3–5 are organized around the systematic practices. In this chapter, we organize the discussion of integrated knowledge exchange around three case studies. We present the experiences of three universities and observe how their *integrated activities* strengthen knowledge exchange.

The domains of education, research, and catalyzing innovations individually make important contributions. The objective of education is to create learning environments and experiences that enable students to develop a deep working understanding of established fundamentals and emerging knowledge, and of essential life and professional skills, including know-how in research and innovation (Chapter 3). Research is about making *discoveries* at the frontier of knowledge, often revealing phenomena or truths that have previously existed but were unknown or unexplained (Chapter 4). Catalyzing innovation produces *creations*—synthesized objects, processes, and systems that did not exist prior to their development at the university. These creations include technologies, inventions, business ideas, medical procedures, and works of art (Chapter 5).

This decomposition into domains and academic practices is a simplification made to allow an orderly presentation. Certainly, there are practices that lie largely within one domain. But in reality, the three domains have significant overlap, as suggested in Fig. 2.1. Many of the important contributions of universities emerge from these overlaps: the speedy introduction of cross-disciplinary and emerging thought into education, the preparation of young entrepreneurs, and the deliberate maturation of technologies to close their readiness gap. The three-way overlap

© Springer Nature Switzerland AG 2020
E. Crawley et al., *Universities as Engines of Economic Development*,
https://doi.org/10.1007/978-3-030-47549-9_6

contains centers that take on societal issues with an intention of direct impact and includes the engagement of students in research.

This integration occurs naturally if the internal boundaries of the university are low and permeable. Crossing boundaries happens as a result of shared goals: education, research, and catalyzing innovation are essentially about creating and exchanging new knowledge. The actions of integration are executed by the same actors: faculty members and students. A conversation can start after a provocative lecture, introduce insight from research, and move seamlessly into sketching out a creation.

The importance of this behavior—overlaps in domains and routine crossing of boundaries—leads to the principle of integration in knowledge exchange introduced in Chapter 2 and summarized in Box 2.2. It emphasizes that it is the cross-disciplinary and integrated activities in education, research, and catalyzing innovation, all engaging with partners, that make knowledge exchange more effective. Following the logic of Chapter 2, this knowledge exchange stimulates innovation and entrepreneurship and contributes to economic development.

In the discussion below, we examine the cases using three questions:

- Do the academic practices appear in the narratives, and thus represent real patterns of behavior at the university?
- Is there integration of the practices within each of the three domains (education, research, and catalyzing innovation) that leads to the outcomes of the domain?
- Is there integration at an overall level—mixing of education, research, and catalyzing innovation—that yields the overall emergent outcomes of the university, knowledge exchange?

To learn about the nature of academic integration, we invited three integrated case studies: the Singapore University of Technology and Design (SUTD), University College London (UCL), and Pontificia Universidad Católica de Chile (PUCC). Told largely in their own words, the cases demonstrate the smooth integration of the 11 practices into a coherent whole. The cases are annotated with the practice icons, indicating how leading universities are deploying the academic practices (Fig. 6.1).

After the case narrative, each of the three case studies is analyzed with a focus within a particular domain and in a particular area of overlap. The SUTD analysis examines integrated education and its overlap with innovation. The UCL case analyzes integration within research and its overlap with education. The PUCC case is examined from the perspective of integrated innovation and its overlap with research. In this way, we discuss integration within each domain and within each area of overlap.

EDUCATION — Chapter 3		Integrated Curriculum
		Teaching for Learning
		Emerging Thought
		Preparing for Innovation
RESEARCH — Chapter 4		Impactful Research
		Collaborative Research
		Centres
		Student Researchers
CATALYZING INNOVATION — Chapter 5		Maturing Creations
		Facilitating Dialog
		Entrepreneurial Venturing

Fig. 6.1 The 11 academic practices with their icons that indicate where in the text the practice is discussed

6.2 Case Study: Singapore University of Technology and Design (SUTD)

The Case of Singapore University of Technology and Design:

A university that is designed to seamlessly integrate education, research, and innovation nurtures graduates who are industry-ready, world-ready, and future-ready.

6.2.1 Advancing Knowledge and Nurturing Technically Grounded Leaders and Innovators at SUTD

If you were to create a world-class university from scratch for the twenty-first century, what would you do? To provide an answer, in 2009 the Government of Singapore invited the Massachusetts Institute of Technology to lead the collaborative efforts to design, develop, and implement the successful opening of a state-of-the-art institution that would become the fourth autonomous university in the country. By May 2012, the Singapore University of Technology and Design admitted its first student class. SUTD aims to lead the fourth industrial revolution by harnessing the best practices and values of the east and the west.

6.2.1.1 SUTD Education

The SUTD academic structure is designed to support a curriculum that develops technically grounded leaders who will contribute to society through technology and design, and make an impact on the world. The academic programs are designed with an outside-in approach that starts with considering industry's needs and delves deeply into the challenges the world faces today. Based on the concept that society needs products, processes, systems, and services, SUTD is not structured via traditional disciplines, but into four pillars (core areas of specialization): Architecture and Sustainable Design, Engineering Product Development, Engineering Systems and Design, and Information Systems Technology and Design (Fig. 6.2). More than 90% of the content of the degrees and undergraduate curriculum was originally developed by MIT.

During the first three terms of undergraduate education, students take common classes together in a cohort-based learning format that fosters collaborative learning and a sense of teamwork, ownership, and belonging. Learning incorporates hands-on activities, such as simulations, demonstrations, and problem sets, where students are challenged to devise solutions in context.

Upon completing the first three terms, students select to specialize in one of the four pillars, graduating with the respective degrees. SUTD's four pillars offer a modern engineering and architectural education that crosses traditional disciplines. In the senior year, students participate in a two-term capstone subject, which requires them to work in project teams with students from at least two pillars, solving real-world challenges, resulting in a truly multidisciplinary education.

Curriculum

		Architecture and Sustainable Design (ASD)	Engineering Product Development (EPD)	Engineering Systems and Design (ESD)	Information Systems Technology and Design (ISTD)
PILLAR	**TERM 4 to 8**	**Capstone Integrated Design Experience**			
			Technical Application Electives		
		Architecture Core	**Product Design Core**	**System Design Core**	**Info Design Core**
		Entrepreneurship, Management, Social Sciences, Economics, Humanities, Arts			
FRESHMORE	**TERM 3**	**Modeling the Systems World, Engineering in the Physical World, The Digital World, Introduction to Biology[1] and introduction to Physical Chemistry[2]**			
	TERM 2	**Advances Math II, Physics II, Introduction to Design and HASS Core Subject**			
	TERM 1	**Advances Math I, Physics I, Chemistry and Biology: Natural World and HASS Core Subject**			

Design Projects   **Advances Pillar Electives**   *Information is subject to change*

[1] Four full credit subjects (or equivalent) per term (x 8 terms)
[2] Half-credit subject

Fig. 6.2 The SUTD curriculum framework

Design is another critical element in the curriculum. In addition to providing strong technical-grounding, and cultivating creativity and a perceptive mind, design is literally everywhere at SUTD. Students are not passive spectators in their own learning; they are engaged in a vibrant design and hands-on culture within and outside the class-room, supported by the Big-Design (Big-D) framework (Fig. 6.3).

PREPARING FOR INNOVATION

There are several integrated practices that foster real-world learning. The 16-week Internship Program is held during the summer vacation each year. Study of the humanities, arts, and social sciences, which makes up more than 22% of the undergraduate curriculum, provides context for design for humanity. Students have the opportunity to explore their own personal interests and activities during the Independent Activities Period.

SUTD is committed to attracting talent from the top universities around the world. The student faculty ratio is 11:1. The SUTD workforce is made up of 52 nationalities. It partners with more than 45 global universities to offer students short-term overseas exchanges, summer programs, internships, and immersion opportunities. Partner universities include Stanford, Berkeley, Yale, Zhejiang, Tokyo Institute of Technology, and the Technical University of Berlin. Currently, about 75% of students have had at least one overseas experience.

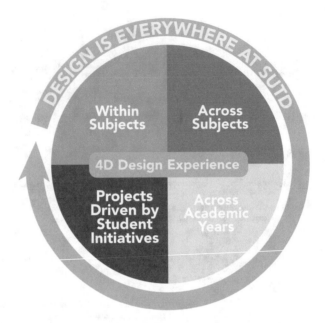

Fig. 6.3 The Four-Dimensional Big-Design (4D Big-D) framework

6.2.1.2 SUTD Research

SUTD has established 13 research centers focusing on operational, economic, and policy research. For example, the Aviation Studies Institute develops advances in air traffic management. The SUTD-MIT International Design Centre is the world's premier hub for technologically intensive design of devices, systems, and services. Other research centers investigate digital manufacturing, advanced robotics systems, cybersecurity, innovative cities, gaming, and industrial infrastructure innovation.

SUTD is also entering its next phase of research growth. It has identified four key areas on which to focus: health care, cities and aviation, supported by capabilities in artificial intelligence/data science. To date, SUTD has fostered more than 800 industry partnerships, which provide students with opportunities for research, internship, and employment.

These research results must be moved to market for commercial impact. SUTD has fostered partnerships with the industry and public research institutions to create testbeds embedded in the living ecosystem around the university. In this way, teams from SUTD are able to move innovations speedily from the ideation phase to proof-of-concept demonstrations. This process also reveals the commercial potential of new innovations.

The Undergraduate Research Opportunity Program (UROP) helps students to effectively apply classroom learning. Students work on cutting-edge research as a junior colleague on a faculty member's research team. Students participate in different phases of research, from writing research proposals and conducting research to analyzing data and presenting results in oral and written form. Students work closely with, and are mentored by, faculty, and partake in various fully funded research projects, which culminate in a final research thesis. UROP helps open up opportunities for students to share and impart their knowledge.

To date, SUTD researchers have published more than 4000 peer-reviewed journal and conference papers and secured more than SGD$350 million in research funding, both from industry, and from various government departments and funding bodies. Based on its research output, SUTD has filed more than 200 technology disclosures and 140 patent applications, and spun off more than 45 start-up companies.

6.2.1.3   SUTD Entrepreneurship

The SUTD Technology Entrepreneurship Program is aimed at nurturing a new generation of innovators through the Lean LaunchPad methodology. Conducted over 4.5 years, students spend a year abroad in the USA and China, immersed in entrepreneurship, and undertaking a work experience term in the Silicon Valley.

The program has four areas of development. The first is STARTsomething, focusing on building a student's entrepreneurial mindset. Then follows BUILDsomething, developing skills for prototyping, building minimum viable products and customer validation. The next phase is LAUNCHsomething, which introduces students to the business practicalities of creating a start-up and running a business. The Entrepreneurship Capstone is a pre-accelerator program where final year undergraduates bring a product from conceptualization to implementation in the market.

Results from SUTD's three graduating classes to date are very encouraging. More than 94% of the students secured employment within 6 months of their final exams with relatively high starting pay. The top hiring sectors included Information and communication, scientific research and development, and finance and insurance. A significant number of students who graduated from the dual degree STEP program founded start-ups within 1 month of their final exams.

The SUTD case is based on input from Professor Chong Tow Chong, President, Singapore, University of Technology and Design.

6.2.2   Analyzing the Singapore University of Technology and Design Case

We organize the discussion of the cases around the three questions stated in this chapter's introduction.

6.2.2.1   Do the Academic Practices Appear in the Narratives, and Thus Represent Real Patterns of Behavior at the University?

Evidence of all 11 academic practices can be found in the SUTD case, as indicated by the icons in the text. For example, reference to patent applications suggest the practice of Facilitating Dialog, and the Entrepreneurship Capstone is an aspect of the practice of Venturing. For SUTD, we look in more detail at the educational practices, examine integration within education, and between education and catalyzing innovation.

The evidence that the educational practices are used at SUTD is summarized in Table 6.1, which maps the practices to the case.

Table 6.1   Evidence of the education practices found in the SUTD case

 CURRICULUM	*Integrated curriculum*—is developed by considering "industry's needs" in the curriculum design, and structuring the education into "nontraditional pillars" that prepare students to design "products, processes, systems, and services."
 LEARNING	*Teaching for learning*—is reflected in the discussion of learning that incorporates hands-on activities such as simulations, demonstrations, and problem sets, where students are challenged to devise solutions in context.
 EMERGING THOUGHT	*Education in emerging thought*—is evidenced by an interdisciplinary curriculum that "crosses traditional disciplines" and requires students to work on projects with students from other pillars.
 PREPARING FOR INNOVATION	*Preparing for innovation*—is demonstrated by the strong evidence of design as a key theme: "design is literally everywhere at SUTD," which engages students "in a vibrant design and hands-on culture within and outside the classroom" preparing future innovators. It reappears in the first three steps of the LaunchPad methodology.

6.2.2.2   Is There Integration of the Practices Within the Education Domain that Leads to the Outcomes of the Domain?

The outcome of education is talented graduates. The education practices are built on Integrated Curriculum, which establishes the learning outcomes and the structure of academic activities. We consider it the foundational practice for education. SUTD chose a curricular structure that "is not structured via traditional disciplines, but into four pillars (core areas of specialization)." It provides a stronger framework to develop the desired outcomes.

An important input to curriculum comes from Emerging Thought. The SUTD pillars allow new cross-disciplinary thought about design to flow into the curriculum. As a result, students work in an approach that "crosses traditional disciplines."

The university's approach to Teaching for Learning complements the learning outcomes defined in the Curriculum. SUTD incorporates into its pedagogical approaches exactly the elements needed to prepare design-oriented students: "hands-on activities, such as simulations, demonstrations and problem sets, where students are challenged to devise solutions in context."

For universities that embrace innovation, there is an additional practice of Preparing for Innovation. This builds on curriculum and learning practices to prepare students as entrepreneurs, managers, and leaders. SUTD extends its educational programs to seamlessly advance the preparation of entrepreneurs through its pervasive focus on design, and its program START something.

6.2.2.3   Is There Integration at an Overall Level—Mixing of Education, Research, and Catalyzing Innovation—That Yields the Overall Emergent Outcomes of the University, Knowledge Exchange?

We use the SUTD case to examine the integration of education with catalyzing innovation. We just discussed an important connection, the flow of skilled entrepreneurs (the outcome of Preparing for Innovation) into Entrepreneurial Venturing, whose outcomes are successful ventures. The two go hand in hand. In fact, the SUTD program START something begins in the domain of education for its first few steps and finishes in the practice of Venturing.

The practice of Facilitating Dialog engages external partners to learn and understand their needs. This informs the desired learning outcomes, which are determined as part of Integrated Curriculum. At SUTD, "The academic programs are designed with an 'outside-in' approach that starts with considering industry's needs and delves deeply into the challenges the world faces today."

The SUTD case demonstrates extensive integration within the practices of education, and between education and catalyzing innovation.

6.3   Case Study: University College London (UCL)

The Case of University College London:

Building on expertise in translational and interdisciplinary research, University College London develops programs of integrated education and open innovation that deliver long-term benefits for a global society.

6.3.1   Integrated Model of Research, Education, and Innovation at UCL

University College London (UCL) is a leading research-intensive university recognized for its radical and critical thinking and its widespread influence. UCL ensures that the academic environment, infrastructure, and administrative processes are designed to promote cross-disciplinary activity. It provides incentives for staff, students, and partners to work together and transform how the world is understood and how global problems are solved.

UCL has identified a set of Grand Challenges that inform the relevance of student learning and research. The currently defined challenges are: Transformative Technology, Sustainable Cities, Global Health, Human Well-being, Justice and Equality, and Cultural Understanding.

Working in partnership with governments, industry, NGOs, and the higher education systems of other countries, UCL is a university that listens, learns, and helps to build capacity. UCL's ambition, expressed in its Strategy, is that by 2034, London will be the global leader in generating jobs, transforming communities, and driving prosperity and societal well-being.

6.3.1.1   UCL Education

The advances in education at UCL are derived from a strong basis of relevant cross-disciplinary research. For example, the UCL Connected Curriculum is built on the idea that intellectually curious students learn better when they directly experience the outcomes of research and connect them with other perspectives (Fig. 6.4). To support such learning, leading scientists visit classes and facilitate discovery opportunities. Students have access to funded summer research positions and participate in projects that may lead to publications. This model suggests that engaging students in research makes learning more relevant and authentic, changing the nature of the dialog between staff and students to become one of partnership.

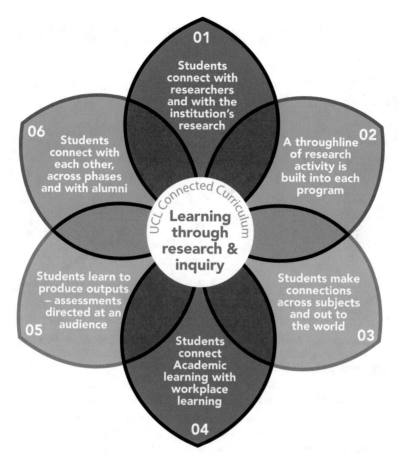

Fig. 6.4 UCL Connected Curriculum: a framework for research-based education to ensure that all students are able to learn through participating in research and enquiry at all levels of their program of study

This ethos is embedded in the Integrated Engineering Programme (IEP), which is built on a common curricular structure across engineering departments (Fig. 6.5). It is an educational framework that combines innovative teaching methods and an industry-oriented curriculum with a discipline-specific, accredited degree program. The IEP supports the principle that engineering is a people-centered, creative discipline that works best when it is diverse, inclusive, and considers social contexts.

Undergraduate students register for a core discipline. The educational program starts with a broad introduction to the challenges facing society that can be addressed by engineering, as well as core disciplinary modules and the first of a thread of instruction and activities in Design and Professional Skills. Through this approach, students

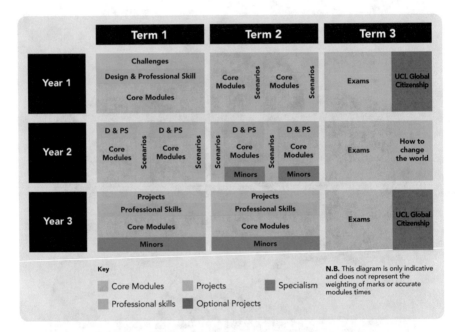

Fig. 6.5 The Integrated Engineering Programme (IEP), an undergraduate teaching framework, emphasizes interdisciplinarity, creativity, and communication alongside teamwork, learning through projects and the social context of engineering

develop the valued workplace and life skills of critical independent thinking, problem-solving, communicating, and working with uncertainty. In the core activities, students learn in a variety of ways, including workshops, flipped lectures, field trips, and experimental or computer labs.

The integrating elements of the IEP are the real design experiences built into each year. In years 1 and 2, students encounter Scenarios. Each Scenarios is part of a 5-week cycle in which students spend the first 4 weeks acquiring relevant skills and knowledge, followed by a 1-week intensive research or design project (Fig. 6.5). Students engage in authentic projects that aim to develop transferable skills and provide an opportunity for students to apply knowledge from the taught program. At the beginning of year 1, students tackle challenges: two 5-week sessions on the themes of sustainability and global health. At the end of year 2, they participate in a How to Change the World capstone humanitarian challenge, an intensive and interdisciplinary studio-based group project with industry and government involvement.

The IEP also includes a compulsory three-module minor to build intellectual diversity. Minors are aligned with UCL's research challenges and use up-to-date insights to design upper level courses. There are currently more than 15 choices of minor topic areas.

6.3.1.2   UCL Research

UCL is a research-intensive university, with a cultural and programmatic emphasis on translational and interdisciplinary efforts. UCL staff work in research domains that are cross-disciplinary communities spanning UCL and partner organizations. They foster higher collaborative impact, and attract major funding, in neuroscience, personalized medicine, populations and lifelong health, cancer, environment, eResearch, collaborative social science, food, metabolism, society, space, and microbiology.

Although each university faculty oversees the disciplinary scope for its own doctoral cohort, the Doctoral School is tasked with aligning research with UCL's 2034 institutional Strategy, and ensuring academics work in a high-quality research-training environment. PhD students are trained with a wide practical skill set that ranges from finance, to negotiation, to coaching style.

UCL provides support to investigators to help them improve their research effectiveness. They can learn how to define objectives, scope, and price of projects; negotiate terms of agreements and manage schedules; and make the connections that lead to further research opportunities and demonstrate impact. A special UCL Venture Research Fellowship supports researchers who explore crucial questions in an open-ended context.

Competitively won research grants and contracts account for more than a third of UCL's income. The main research funding comes from the portfolios of the UK Research and Innovation, the Engineering and Physical Science Research Council, and the Science and Technology Facilities Council, and from a diverse number of organizations including the Higher Education Innovation Fund. For researchers, there are multiple sources of funding that support connections beyond academia and help to apply research in industry. Options include arranging a secondment, setting up a business partnership, or hosting an industry expert-in-residence.

The UK Research Excellence Framework, with its emphasis on impact, plays a new, pivotal role in university research. Although there is now more emphasis on directed research, UCL's high caliber researchers excel at delivering high-impact outputs, producing top publications and fulfilling short-term tasks. They do this through their significant experience working with industry, and the stellar talent pool derived from a meticulous hiring process.

6.3.1.3   UCL Innovation

Innovation and Enterprise is the function at UCL that offers a wide range of support, training, and advice, helping to turn knowledge and ideas into solutions that benefit society. It assists staff and students in commercialization, knowledge transfers partnerships, spinouts, and intellectual property licensing. And it helps organizations to better access UCL intellectual assets. For example, UCL Consultants Ltd. was set up to draw on UCL's research competence to provide academic consultancy services, and bespoke short courses for business. UCL Knowledge Transfer Partnership contributes to solving strategic challenges by engaging world-leading academics with counterparts in industry and enterprise.

UCL is involved in several large strategic relationships. It is working with Facebook, Amazon, and Google to ensure AI technology will address challenges such as climate change, food security, wealth inequality, lifelong health, and progressive education. It also is collaborating with Barclays to play a leading role in applying technology to the practice of the law.

UCL supports individuals and small businesses in incubators, accelerators, and post-accelerators and provides co-working spaces and sector-led virtual communities. UCL has assisted many London's start-ups. It also works in partnerships: for example, with Cisco and EDF, and it sponsors the Innovation Hub IDEALondon, creating the future of the Internet of Things. In addition to multiple venture fund initiatives, in partnership with NHS Trusts, UCL has developed UCL Business, comprising 247 patent families and 75 equity holdings, as the commercialization arm to bring to market truly world-changing technologies.

There are a number of efforts aimed at building entrepreneurial capabilities and igniting start-ups. The UCL Enterprise Fellowship Programme provides training for postdocs and academic staff to develop an entrepreneurial mindset and create impact when engaging with organizations outside of academia. Conception X is a new  business accelerator for PhD students to successfully transit from thesis to deep tech start-ups by providing mentorship, network, and initial funding.

The UCL case is based on input provided by John Mitchell, Vice-Dean of Engineering (Education); Anthony Kenyon, Vice-Dean of Engineering (Research); and Jane Butler, Vice-Dean of Engineering (Enterprise), University College London.

6.3.2 Analyzing the University College London Case

6.3.2.1 Do the Academic Practices Appear in the Narratives, and Thus Represent Real Patterns of Behavior at the University?

Again, evidence of all 11 academic practices can be identified in the case narrative, as suggested by the icons in the case text. We use the UCL case to examine the integration within the research practices, and between research and education. Table 6.2 offers short quotations from the UCL case that substantiate the use of the research practices.

6.3.2.2 Is There Integration of the Practices Within the Research Domains that Leads to the Outcomes of the Domain?

The outcomes of research are discoveries, often revealing phenomena or truths that have previously existed but were unknown or unexplained. We consider Impactful Fundamental Research as the foundational practice of research. It

Table 6.2 Evidence of the research practice found in the UCL case

	Impactful fundamental research—is discussed in the context of the UK Research Excellence Framework: "UCL's high caliber researchers have excelled at delivering high-impact outputs, producing top publications, and fulfilling short-term tasks."
	Collaborative research within and across disciplines—is evident from the emphasis on cross-disciplinary research: "staff work in research domains that are cross-disciplinary communities spanning UCL and partner organizations."
	Centres of research, education, and innovation—are strongly developed: "UCL is working with Facebook, Amazon and Google to ensure AI technology will address challenges such as climate change, food security, wealth inequality, lifelong health and progressive education."
	Undergraduate and postgraduate student researchers—is part of the experience: "Students have access to funded summer research positions, and participate in projects that may lead to publications. … engaging students in research makes learning more relevant and authentic …"

charges researchers to do fundamental research along a spectrum from curiosity-driven to use-inspired. At UCL, this span is indicated by a balance of traditional measures of scholarship alongside "more emphasis on directed research."

Impactful Fundamental Research sets the direction of discovery, while Collaborative Research is an approach, so they go hand in hand. Collaboration based on earlier Fundamental Research can "foster higher collaborative impact and attract major funding." The small scholarly core team that is typical of collaboration makes "connections that lead to further research opportunities and demonstrate impact," including the development of Centres of

Research, Education, and Innovation. Of course, the intellectual foundation of Centres often rests in Impactful Fundamental Research.

6.3.2.3 Is There Integration at an Overall Level—Mixing of Education, Research, and Catalyzing Innovation—that Yields the Overall Emergent Outcomes of the University, Knowledge Exchange?

We use the UCL case to demonstrate integration between research and education. Researchers play a key role in strengthening education. They develop new and interdisciplinary Emerging Thought and move it quickly to the curriculum. This is a common pattern at UCL, where "the UCL Connected Curriculum is built on the idea that intellectually curious students learn better when they directly experience the outcomes of research and connect them with other perspectives."

Another important integration of education and research is sharing a precious resource: students. Through education, students gain knowledge and skills. In research, they develop research know-how and gain frontier research knowledge. "Students have access to funded summer research positions, and participate in projects that may lead to publications," suggesting that "engaging students in research makes learning more relevant and authentic." As a result of research links to education, students leave the university prepared for a career of knowledge exchange.

Clearly, at UCL, there is significant integration within research, and between research and education.

6.4 Case Study: Pontificia Universidad Católica de Chile (PUCC)

The Case of Pontificia Universidad Católica de Chile:

A transformative program prepares a new generation of technology innovators, develops pioneering applied research outcomes, and fosters entrepreneurial capacity, thanks to a globally connected environment and a world-class organization.

6.4.1 The Clover: A Transformative Program for an Innovation-Based Economy at PUCC

Chile aims to become the first Latin American country where economic and social development are built upon technology innovation. This transformation requires a collaboration among government, industry, academia, and civil society. The Pontificia Universidad Católica de Chile (PUCC) is playing a pivotal role as the leading engine for driving the country's technology research base, aligning the national research and development culture towards an entrepreneurial mindset with deepening connectivity to the world's foremost technology-innovation ecosystems. The lucky four-leaf clover was chosen to represent this ambitious transformation program, which is based on four pillars: education, applied research, entrepreneurship, and an enabling environment. At the heart of this transformation is PUCC's engineering school: PUC Engineering.

6.4.1.1 PUC Education

PUC Engineering developed a suite of new courses, activities, and programs across both the formal engineering curriculum and the co-curricular space. Learning experiences are designed to support and guide students along the "entrepreneurial pipeline," from initial social awareness-raising and engagement with the concept of technology-driven entrepreneurship, to the commercialization of a new product or system.

Opportunities for students to learn about entrepreneurship are available through the "visible and invisible curriculum" (Fig. 6.6). The visible curriculum's mandatory courses are taken by all engineering students in fields such as engineering design, entrepreneurship, and innovation. In the invisible curriculum, co-curricular activities support progressive development of skills in three distinct areas: science and technology, information technology and applications, and social entrepreneurship. Entrepreneurial activities also include a competition that showcases the school's best student projects, and an opportunity for PUC Engineering students to spend 1 month in vibrant ecosystems, such as Silicon Valley and Boston.

Entrepreneurship learning touches each student at least twice. In the first year, the cornerstone course, Engineering Challenges, introduces first years to engineering design practices around a relevant social concern by following a user-driven design process. In this project-based course, students work on teams of six or seven to design devices that tackle a real-word problem of social relevance, such as improving resilience to extreme natural hazards in Chile or adapting to climate change.

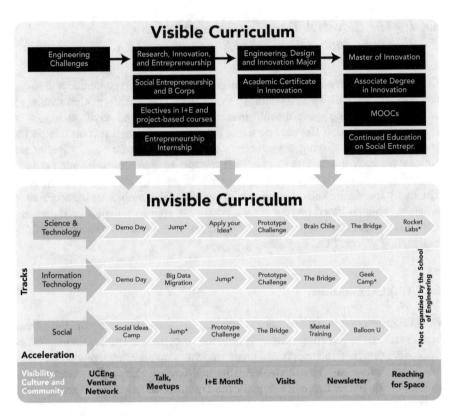

Fig. 6.6 PUC Engineering's *visible and invisible curriculum*, which raises students' awareness of the entrepreneurship opportunities open to them, many of which are voluntary

In the third year, the course Research, Innovation and Entrepreneurship (RI&E) provides a hands-on innovation experience for about 400 engineering students per semester. Developed in partnership with the Centre for Entrepreneurship and Technology at the University of California Berkeley, the course challenges cross-disciplinary student teams to develop technology-based solutions to key problems facing Chile. Its aim is to familiarize students with the complexity and uncertainty encountered in the development of an innovative product or service. A second objective is to develop the skills to recognize the technology business opportunities necessary to taking the product or service to the market and, thus, meet societal needs.

PREPARING FOR INNOVATION

At the midpoint of the RI&E course, student teams develop a functional prototype. By the close of the 16-week course, students have developed an innovative technology-based solution, and they pitch their idea to a panel of entrepreneurs, professors, and industry experts. Support by Chilean entrepreneurs throughout this course plays a critical role in the successful delivery and achievement of the desired competencies. The entrepreneurs make presentations about their business experiences, and mentor the teams.

6.4.1.2   PUCC Applied Research

IMPACTFUL RESEARCH

The transformed research program aims to produce pioneering applied research that will support innovation that transforms lives. A distinct element of the new practice is the Seed Fund program, which scales up the impact of applied research at PUC Engineering through strategic partnerships on cutting-edge topics with universities such as the Massachusetts Institute of Technology, the University of Edinburgh and Texas A&M. Student engagement is designed as a stepping-stone towards a deeper strategic research collaborations with global leaders in different fields, introducing cutting-edge knowledge into students' education. The Seed Fund program also provides a platform from which PUC Engineering is able to access state-of-the-art infrastructure and technology, explore new global networks, and access international sources of research funding.

EMERGING THOUGHT

Each Seed Fund project is led by two faculty members, one from PUC Engineering and one from the partner university. In addition to the joint research, the fund supports exchange visits by both the faculty leads and the undergraduate and postgraduate students engaged in the study. Significant outcomes are expected from the collaborating team, including joint high-impact publications, and a next-phase collaborative proposal for broader external funding. For example, in 2016 one faculty member secured a collaborative research Seed Fund grant with Harvard University in the field of astronomy and data science. The research was focused on developing a platform to process, analyze, and catalog the data produced by Chile's new telescopes, as well as developing tools to visualize the results. The fund enabled the research team to have access to state-of-the-art equipment and to collaborate with leading international experts in the field.

COLLABORATIVE RESEARCH

PUC Engineering is in the process of establishing new high-impact applied research centers in critical areas of national and global concern, such as health, sustainability, information, and engineering in science. Collaborative and interdisciplinary by nature, these new centers will draw on and build upon the school's growing strengths in applied research and engineering design. They will drive state-of-the-art applied research that cuts across the traditional engineering disciplinary boundaries. Dedicated managers will be appointed at each center to build research capacity in the field and foster collaboration across and beyond UC Engineering.

CENTRES

This research program supports economic and social development built upon technology innovation in two ways. First, it supports development through the direct economic value created by a number of new start-up companies that have flourished through students and professors transferring their research. A good example is Amazon's investment in the Not Company, aimed to design new food products based on vegetable proteins that resemble milk. This company, cofounded by a computer science professor, is based on the use of highly effective machine learning algorithms.

A second path is the formal development of contract research with industry and government through a new model of PUCC's Industry Liaison Office, which combines the best academic capacities of professors and students around a challenging topic. Being a link with industry and the government, this office is dedicated to strengthening and catalyzing knowledge transfer. Therefore, it is a gateway through which PUC Engineering addresses a wide range of societal issues.

Another interesting aspect of PUC Engineering's is its Undergraduate Research program, which allows undergraduate students to become actively involved in research projects of Faculty members. Students interested in this program can browse the research opportunities uploaded by faculty members on a web-based platform. Since the inception of the program, the number of published  papers co-authored by undergraduate students has increased significantly.

6.4.1.3   PUCC Entrepreneurial Culture and Capacity

During the earlier stages of the transformation, PUC Engineering was focused on consolidating global academic innovation partnerships, redesigning professional education programs, and bridging PUCC with Silicon Valley. Then, in 2015 Brain Chile was created as a competition to fill the funding and support gap for emerging  research-based start-ups between developing a proof-of-concept or prototype and securing seed-funding investment.

Brain Chile was originally designed to support innovations emerging from research at PUC Engineering. However, the initiative quickly established a national scope due to the growing interest from all stakeholders across Chilean higher education, and Santander became a sponsor. By its second iteration, 135 projects from across the country applied to participate in the program. Eighty percent of participants were affiliated with a Chilean university, with the remaining 20% coming from Chilean business or the entrepreneurial sector. Today, Brain Chile has an international scope, with 238 participant teams from universities in

Latin America (24%), Santiago (45%), and other Chilean regions (31%), marking a five-fold net growth since its first version.

Brain Chile encompasses three key phases. In the first phase, all applicants receive online training in business and entrepreneurship concepts. In the second phase, 36 teams are selected to participate in a Braincamp, an intense week of workshops on business models, prototyping, intellectual property, pitching, and industry mentoring. Braincamp ends with 12 teams reaching the third phase, known as the Acceleration phase, which includes a customer identification process, coached by PUCC instructors who have been trained on NSF's I-Corps Lean Start-up methods. In parallel, teams receive seed funds, mentoring, design and IP consulting, and pitch training, so finalists become prepared to compete for cash prizes from an investor panel of venture capitalists.

6.4.1.4   PUCC Environment and World-Class Organization

To support this transformative program, PUCC is implementing an Academic Intelligence System to continuously monitor the school's operation and performance. It is establishing an International Liaison Office to enhance global university partnerships and host international visitors on campus. It is building new facilities; for example, it launched the Fabrication Laboratory at the PUC Center of Innovation and the Collaborative Laboratory in Microelectronics, Sensors, Control, and Engineering Design.

PUC Engineering also led the creation of the first two new interdisciplinary institutes within the university; one in biomedical engineering, and another in artificial intelligence. It launched dual or joint doctoral programs with universities such as the University of Edinburgh and Kings College London in the UK, and Notre Dame in the USA. To develop its team, it established a mentorship program for early career academics and implemented a new promotion and biannual faculty evaluation process, which includes specific indicators to account for academic contributions in innovation and entrepreneurship.

This case was contributed by Juan Carlos de la Llera, Dean, School of Engineering; Jorge Baier, Associate Dean of Engineering Education; Isabel Hilliger, Associate Director Assessment and Evaluation; Constance Fleet, Associate Director for Innovation, School of Engineering; Pontificia Universidad Católica de Chile.

6.4.2   Analyzing the Pontificia Universidad Católica de Chile Case

6.4.2.1   Do the Academic Practices Appear in the Narratives, and Thus Represent Real Patterns of Behavior at the University?

In the PUCC case, all eleven of the academic practices are deployed, indicated by the icons in the case text. In this case, we seek evidence of integration within catalyzing innovation, and between innovation and research. The evidence of the practices of catalyzing innovation that are employed at PUCC is summarized in Table 6.3.

6.4.2.2   Is There Integration of the Practices Within the Domain of Catalyzing Innovation that Leads to the Outcomes of the Domain?

The outcomes of catalyzing innovation are creations—synthesized objects, processes, and systems that have never existed prior to their development at the university. While this potentially spans a broad spectrum from art to business, we focus on the creations arising from science, technology, and entrepreneurship.

Maturing Discoveries and Creations is the foundational practice for catalyzing innovation. It identifies new technologies in the context of markets, and matures their technology and business readiness. Maturing Discoveries is tightly coupled to

Table 6.3 Evidence of the practices of catalyzing innovation found in the PUCC case

	*Maturing discoveries and creations*—is reflected in Brain Chile that "was created as a competition to fill the funding and support gap for emerging research-based start-ups between developing a proof-of-concept or prototype, and securing seed-funding investment."
	Facilitating dialog and agreements—is manifest in "the formal development of contract research with industry and government through a new model of an Industry Liaison Office," And the establishment of an "International Liaison Office."
	University-based entrepreneurial venturing—is evident in the research program which "supports development through the direct economic value created by a number of new start-up companies that have flourished" and brought new products and systems to markets.

the practice of Facilitating Dialog and Agreements, which often surfaces market and business needs (the market pull). At PUCC, this sometimes leads to engagements, "which combines the best academic capacities of professors and students around a challenging topic." The licensing of new inventions also links these practices.

There is a parallel relationship between Maturing Creations and Entrepreneurial Venturing. Maturing Creations produces technologies that can be disseminated (tech push) to Venturing, so that at PUCC Brain Chile was designed to "support innovations emerging from research at PUC Engineering." Venturing can exert market pull on Maturing Creations. Facilitating Dialog and Agreements oversees licensing to Ventures.

6.4.2.3 Is There Integration at an Overall Level—Mixing of Education, Research, and Catalyzing Innovation—that Yields the Overall Emergent Outcomes of the University, Knowledge Exchange?

The principal interactions between research and catalyzing innovation is the flow of discoveries and knowledge from Fundamental Research, Collaborative Research and Centres into the practices of innovation including Maturing Discoveries and Creations, Facilitating Dialog and Entrepreneurial Venturing. At PUCC, this occurs when the outcomes of "pioneering applied research" and Seed Fund projects find their way into the "entrepreneurial pipeline" and to industry with the help of the Industry Liaison Office. PUC can observe that companies "have flourished through students and professors transferring their research."

The other form of integration between innovation catalyst and research is when the needs of society and the market, as understood by entrepreneurs and industry liaisons, are transferred to researchers. For example, this occurs when industry sponsors research, or when students participating in Braincamp identify needs they pass back to their research supervisors.

The PUCC case demonstrates integration within catalyzing innovation, and between innovation and research.

6.5 Summary and Observations on the Integrated Cases

The outcomes of a university—talented graduates, research discoveries, and innovation creations—emerge from the whole fabric of the university. With increased emphasis on knowledge exchange, these outcomes will intensify innovation in the economy and lead to sustainable economic development.

It is clear from this small sample that ambitious universities seeking to energize innovation are deploying the academic practices in a systematic manner. We found that PUCC, SUTD, and UCL each report patterns of behavior that align with all 11 practices.

By examining the cases, we observe that the practices within each of education, research, and catalyzing innovation constitute an integrated and interaction assembly: curriculum is meaningless without learning; impactful research and collaborative research go hand in hand; and entrepreneurial venturing is enhanced by licensed and matured technologies.

Likewise, the practices in the three domains are integrated: research stimulates curriculum reform, well-prepared students strengthen venturing, and dialog helps to guide research. It is useful to break the behavior of the university into practices to explain them and provide references that others can adapt, but in reality the practices work together, as suggested by the principle of Integration in knowledge exchange, presented in Box 2.2.

Chapter 7
Supporting the Academic Mission of the Adaptable University

7.1 Introduction

7.1.1 The Adaptable University Supports the Academic Practices that Contribute to Innovation

In the first half of this book, we discuss the significant ways in which universities contribute to economic, social, and cultural development (Chapter 1). Because of high expectations and investments, we focus on the contribution of the university to sustainable economic development, while acknowledging the parallel importance of university contribution in the social and cultural spheres. We highlight the approaches that can generally be adapted by research universities seeking to embrace innovation. These include technical universities and comprehensive universities with science and technology programs.

Chapter 2 identifies knowledge exchange as the enabler of economic development, and the university's broader engagement with society. The actual impact is made through education, research, and catalyzing innovation, all engaging in knowledge exchange. These core academic activities are presented in Chapters 3–5. Their integration is discussed in Chapter 6.

The second half of the book casts the university as an adaptable organization, able to evolve and ready to support the expanded academic mission in innovation. The adaptable university has porous boundaries that allow deep engagement with partners. It runs effectively, and its supporting practices encompass stakeholder engagement, culture, strategy, governance, facilities, and human resources. These supporting practices are discussed in this chapter.

© Springer Nature Switzerland AG 2020
E. Crawley et al., *Universities as Engines of Economic Development*,
https://doi.org/10.1007/978-3-030-47549-9_7

> **Box 7.1 Objectives of Supporting Practices**
>
> The general objective of the supporting practices is to enable and strengthen the university's academic activities in education, research, catalyzing innovation and knowledge exchange, and to create a more supportive intellectual ecosystem in which faculty, staff, and students can flourish and contribute.
>
> These practices increase the coherence of the university, make it more adaptable, focus its resources on important tasks, and allow more effective communication internally and externally.

Viewed through the lens of economic development, the specific objective of these practices is to more effectively support the academic programs of the university and create a more supportive intellectual ecosystem, all while allowing it to adapt to its expanded mission in innovation.

7.1.2 The Supporting Practices

Based on our experience, we have chosen a set of supporting practices that are most important in supporting the academic programs (Table 7.1). These practices support continuous improvement at existing universities and the creation of new universities. As with the academic practices, each university will have to find its own way to adopt these supporting practices. The practices are discussed below, illustrated by eight case studies.

Table 7.1 The practices that support the academic programs

Practice name	Description of the practice > and its outcome
Engaging Stakeholders	Engaging with and learning the needs of the expanded set of stakeholders associated with economic development > leading to strategy and programs more likely to address stakeholder and university needs
Evolving Culture	Evolving the university's culture and values > encouraging the activities leading to economic development
Mission and Strategic Planning	Revising the university's mission, strategy, and priorities > focusing on its investment of resources, and communicating how the university will contribute and distinguish itself, including in innovation
Governance	Updating decision-making, policies, organizations, and budgets > promoting a clear definition of devolved authority and the transparent decisions necessary to support the strengthened role in innovation
Faculty and Staff Resources and Capabilities	Recruiting and developing high-quality faculty members and professionals > yielding a university community better able to undertake the tasks and programs linked to innovation
Academic Facilities	Providing functional, flexible facilities—Classroom, social spaces, laboratories, workshops, and IT connectivity > enabling new activities in learning, innovation, and collaborative research with industry

All universities have some form of these supporting practices, in a general sense, and some universities have developed them to a high degree. But the practices of Table 7.1 are written to emphasize how the university will adapt to an expanded mission in economic development. Therefore, they are important mechanisms for enabling change.

Three of the practices—strategy, culture, and governance—resonate with John Van Maanen's analysis of the drivers of any organization [1, 2]. His three-perspective model includes: action following strategy based on universal rationality, action following influence and power, and action following habit or culture, which may be neither rational nor visible. These modes of action are always at work, and the three associated practices help to build a university that can adapt more readily to its new missions.

The other three supporting practices play a more executive role in the university's transformation. Engaging with existing and new stakeholders will reveal how the university can help them, and how they can help the university. Faculty and staff development influences how effectively the academic team can contribute. Suitable facilities enable new activities.

7.1.3 The University with Porous Boundaries

Basic here is the notion of the porous university, where there is creative and constructive engagement and movement across boundaries (Fig. 1.1). The main resources that flow into the university are people, facilities, and funding. Low boundaries facilitate stakeholder engagement. These inputs flow to the academic practices in education, research, and innovation, which operate, moderated by culture, strategy, and governance. The outcomes are talented graduates, discoveries, and creations that are exchanged with industry, enterprise, and other partners, universities, and governments (Chapter 2). At each point of contact, low and porous boundaries will better support the university mission.

The transformation that would result from the adaptation of these supporting practices is indicated by the change from the reference situation (our estimate of the current average state of a good university) to the aspirational situation shown in Table 7.2.

7.2 Engaging Stakeholders

7.2.1 The Broader Set of Stakeholders of the Expanded Mission

For the university to adapt and prosper, it is important to engage with stakeholders and learn how their needs can be met by the university. By reflecting on these needs, rationalizing them and reflecting them appropriately into university goals, strategy,

Table 7.2 Reference and aspirational situations for the supporting practices

Reference situation	Aspirational situation
Dialog with government and industry to inform them of the progress of the university	Broad dialog with an expanded set of stakeholders and an understanding of their needs addressable by the university [Stakeholder]
A culture of education and research, a strategy developed by university leaders, and a governance system working effectively	An evolved culture, an inclusive strategy, and revised and transparent governance, all better supporting the expanded role of the university [Culture, Mission, and Governance]
Faculty who are excellent researchers and dedicated teachers	Faculty and staff hiring and development are shaped to strengthen the capacity of the university for economic development [Faculty and Staff]
Good research laboratories and classroom facilities	Learning space, labs, facilities, and IT infrastructure to supports all modes of learning, research collaboration, and innovation [Facilities]

and programs, the university will bring value to the stakeholders. That will be good for the university, as it reflects the contributions of the university to society. It will also garner the recognition and resources the university needs.

Box 7.2 Goals of Engaging Stakeholders

Universities will more effectively support their academic programs and create a more supportive intellectual ecosystem, all while adapting to their expanded mission in innovation,

By engaging with and learning the needs of the expanded set of stakeholders associated with economic development,

Leading to strategy and programs more likely to address stakeholder and university needs

---

Stakeholders are those that have a stake in the university. They can contribute to the university and benefit from it. A university's stakeholders usually include: students as the first priority; faculty and administrators; government as an educational policy maker, regulator and funder; industry; small and medium enterprise; peer universities; philanthropies; and the public. These are shown in the left column of Table 7.3. We use the term partners to identify the subset of external stakeholders with whom the university directly exchanges knowledge (primarily industry, enterprise, and government).

Universities seeking to broaden their mission should deal with more stakeholders in deeper relationships [3]. Small and medium enterprise and industry are now more prominent stakeholders [4]. Government institutions also act as organizational stakeholders. These can include national laboratories, mission departments, and public services (e.g., NASA, ministry of defense, national rail). Public support

Table 7.3 A simplified template of outcomes to and from key university stakeholders

Stakeholder	Stakeholder outcomes that meets university needs	University outcomes that meets stakeholder needs
Students	Attendance, enthusiasm, tuition fees	Education, experience, diploma
Faculty	Teaching, research, innovation	Opportunities, recognition, compensation
Government	Policy, regulation, funding	Contributions to society, economy
Industry	Joint projects, financial and political support	Talent, creations, discoveries
Enterprise	Partnerships, financial support	Talent, creations
Peer universities	Talent, faculty, practices	Talent, faculty, practices
Philanthropies	Funding	Contributions to society, economy
Public	Political support	Insight, health, employment

becomes more important because major research and innovation developments need the trust and confidence of the public.

The view of stakeholder interaction of Fig. 7.1 shows the sense of a circular flow of outcomes in an extended network. Talented students, discoveries, and creations flow from the university to the commercial sector, which produces goods, services, and systems that benefit the broader public. The public supports government, which provides the core funding and policy support to the university.

Reasoning about this network of university stakeholders makes evident the importance of engagement [5]. Engagement is essential, particularly in formulation of mission and strategy. The network identifies the boundaries across which knowledge must be exchanged. It can reveal many of the issues in stakeholder management that face university leaders. In particular, it can help clarify the many, and sometimes conflicting, stakeholder needs. In this chapter, we consider how the university can develop its side of the practice of stakeholder engagement; Chapter 9 considers the alignment by external stakeholders.

7.2.2   External Stakeholder Identification and Engagement

The process of stakeholder engagement begins by identifying the important stakeholders of the university [6]. In the case of economic development, key partners include industry, small and medium enterprise [7], and government institutions [8]. Relevant partners will may be local, global, or sometimes both.

Engagement is about understanding needs. The university learns about the needs of the stakeholder. The university also identifies its needs that the stakeholder can meet. Table 7.3 offers a simplified template of needs of the university and the expanded set of internal and key external stakeholders. One way to learn about needs is by reading stakeholder literature, such as plans, statistics, and media reports. But person-to-person interviews with current leaders and other representative thought leaders is essential.

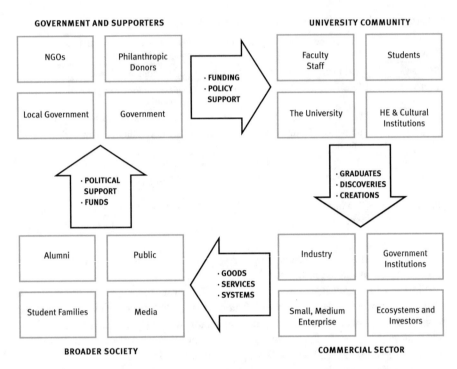

Fig. 7.1 The sense of a circular flow of outcomes in a university's stakeholder network

Engagement is also about subsequent actions [9]. Stakeholder needs can influence the university's mission and programs and are important in evolving culture and updating governance. Stakeholder understanding can lead to new agreements with the commercial sector and to broader involvement by society in the university. An understanding of stakeholders can be instrumental in dealing with government, and with other policy and funding supporters.

The process of engaging stakeholders at a new university is presented in Case 7.1: Skoltech—Designing a University to Deliver Value to Stakeholders. From its beginning, Skoltech was immersed in a developed intellectual community and a commercial ecosystem with many needs. In this complex stakeholder system, a comprehensive stakeholder analysis pointed the way towards satisfying stakeholder expectations.

Case 7.1 The Skolkovo Institute of Science and Technology (Skoltech)—Designing a University to Deliver Value to Stakeholders

Analyzing a stakeholder network identifies important stakeholders, as well as the outputs of the university that are important to them, and suggest strategies for developing stronger stakeholder relationships.

Skoltech was founded in 2011 as a graduate university of science and technology with the mission of accelerating innovation by building bridges between the scientific base and commercial community. It was as a key element of the Skolkovo Innovation Center, a massive economic development program led by the Skolkovo Foundation on behalf of the Russian government.

This web of organizations, combined with general geopolitical factors, presented an extremely complex stakeholder environment. Therefore, an intense effort was launched to identify and characterize the stakeholders, and how Skoltech would create value for them. This effort spanned months, involved document analysis and interviews, and applied quantitative methods.

The first outcome of the effort was the identification of the five key stakeholders.

- The incumbent Russian Federation (RF) Institutions, including universities and the Academy of Science.
- The government of Russia, at regional and national levels.
- Industry, which spans from large and state owned companies to start-ups.
- The Skolkovo Foundation (SkF), the foundation charged with development.
- Skoltech, the central entity of the network for these purposes.

This first phase revealed several unanticipated stakeholder groups, such as the Academy of Science and regional government. It allowed the founding team to reason qualitatively about how to engage with the various parties, and the parties' roles in the network.

A more detailed and quantitative network map was developed for these five stakeholders (see Fig. 7.2). This map identified the important needs of each stakeholder, and where in the network a flow of value is created that meets the needs. The width of the connecting arrow represents the intensity of need. The

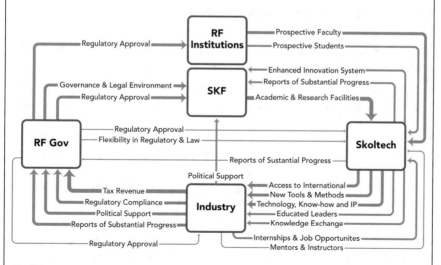

Fig. 7.2 Skoltech stakeholders and the non-monetary flows of value among them

theory of stakeholder networks suggests that the most important outputs of a central entity are the ones that propagate through the network and eventually return to the central entity what it needs. Therefore, Skoltech should act to prioritize the outputs most valued by its stakeholders, to the extent that those stakeholders can return benefit back to Skoltech.

Examining the figure, we can see the most important loop by following the strongest bold lines in the figure. It starts at Skoltech providing to industry "access to international workforce," and "new tools and methods"; industry passes "financial" and "political" support to government; which passes a "governance and legal environment" to the Skolkovo Foundation; which passes "facilities" (and funding) to Skoltech. The importance of this loop is that it focuses attention on the urgency for Skoltech to produce talented graduates and innovations for industry, in contrast, for example, with achieving scholarly recognition.

It became clear that "crossing bridges" was not just a metaphor in the mission, but an essential activity. Other important learnings were that industry and particularly government had an unrealistically short timescale of expectation—so both speed and managing expectations were important. Unlike initially thought, we learned that the existing top universities were not opponents, but supporters—they saw Skoltech as a pathway for their graduating students and as a model for change. As a private university, Skoltech had not expected the influence of the government though the regulatory mechanism of program accreditation—which eventually became a dominant consideration in curriculum design.

7.2.3 Stakeholder Involvement and Influence

Stakeholder engagement takes place at the university level, but also at the level of schools, departments, centers, and programs. For example:

- Designing curriculum by considering the interests and inputs of employers. This is often done by advisory groups made up of faculty, employers, and students (Chapter 3).
- Planning research by consulting with those interested in the outcomes. In performing impactful research, mechanisms for engagement are of great value (Chapter 4).
- Developing innovation programs by cooperating with local entrepreneurs, enterprise, and close industrial partners (Chapter 5).

Another important role of stakeholders occurs when programs are periodically reviewed. While colleagues from peer universities typically perform such reviews, involving industry and enterprise will emphasize the role of economic development. External stakeholders can also become involved in university standing advisory and decisional bodies (Governance).

While we argue for stakeholder engagement, there is no doubt that it can generate tensions. These must be resolved by strong principled governance, strategic planning, and often on a case-by-case basis. For example, including employers in the design of the curriculum can help identify the skills of contemporary practice, but raises the specter of whether graduates will be too narrowly prepared for a life in which career change is common. Industry engagement in research increases knowledge exchange, but can cause tensions due to differences in focus, timescales, and rewards. Government engagement on the topic of economic development can become a hotly contested balance of autonomy and accountability. Not all stakeholders are equally important to the university. Priority should be given to the stakeholders on which the university most relies [10].

When stakeholder engagement is poorly handled, these fears will be justified. Handled openly, with trust and concern for helping society, stakeholder engagement will strengthen core values. These tensions and conflicts are considered partly in the sections on Culture and Governance in this chapter, in Chapter 9 on Stakeholder Alignment, and in Chapter 10 on Change.

7.3 Evolving Culture

7.3.1 Evolving Culture to Support the Expanded Mission

Promoting knowledge exchange across porous boundaries and being more adaptable to the needs of society, require all of the supporting practices. Most important is an evolution of culture [11].

Box 7.3 Goals of Evolving Culture

Universities will more effectively support their academic programs and create a more supportive intellectual ecosystem, all while adapting to their expanded mission in innovation,

By evolving the university's culture and values
Encouraging the activities leading to economic development

Culture is the most powerful determinant of behavior in any organization [12]. As popular wisdom says: culture eats strategy for breakfast. It is a complex combination of customs, habits, collective knowledge, modes of discourse, and attitudes of faculty, students, and staff. Culture is commonly understood by reflective members of the community [13, 14]. It can sometimes be expressed in terms of underlying values or principles [15, 16]. Universities have subcultures that must be considered, and sometimes are in tension: faculty, administration, and students. Subcultures can be distinct by unit and discipline as well.

Culture preserves what is tried and tested and can be suspicious of hasty change, but university culture does evolve. Early universities emphasized teaching, and they catered to a small number of students. Every change since has been debated and contested—growth in student numbers, change from teaching only to the inclusion of research, the concept of measuring impact, and funding by industry. But universities have absorbed these changes and their culture has evolved. The changes associated with emphasis on economic development will also be hotly debated, but in the end the evidence will lead to positive evolution.

Long-standing institutions have developed their culture over generations [17]; that is what makes them distinct and contributes to their success. Even so, a wholly new institution can establish a strong culture with surprising speed. Changing the culture of established institutions or creating a distinct culture in new institutions is a challenging endeavor, but it can be accomplished. In fact, cultures that do not adapt will die.

What is needed is evolution in culture, not radical change. The concepts of accelerating innovation and knowledge exchange across porous boundaries are not narrow or short term. They depend on traditions built up in the past. An evolved culture can inspire and motivate students and faculty while enhancing contributions to innovation and helping to justify significant public investment.

Case 7.2 on The Culture of MIT exemplifies how a university can develop a culture that supports innovation. But as the case discusses, while elements of this culture were present at the founding of MIT, the current culture is a result of several major rounds of evolution.

Case 7.2 The Massachusetts Institute of Technology—The MIT Culture

The culture of a university can evolve and support a shift towards an expanded mission in innovation.

In 2013, MIT conducted a study to identify the key attributes its contemporary culture. This was done to support MIT's institution-building partnerships, so that those involved, be they from MIT or elsewhere, are aware of the organizational culture that underlies MIT.

MIT certainly has a strong and pervasive culture. Larry Bacow, president of Harvard University, says in the introduction of Mens et Mania [1]: "Many countries and many institutions have tried to replicate the success of the world's foremost institution for science and technology. However, MIT is more than just a collection of buildings on the Charles River populated by brilliant students, faculty and staff. MIT represents a particularly unique and often poorly understood culture. Anyone who seeks to replicate or simply understand MIT must first try to understand and appreciate a culture that celebrates quirkiness, choice, independence, entrepreneurship, focus, creativity and intensity."

In the 2013 study, the attributes of MIT's contemporary culture were identified through a review of literature on MIT and other universities, and 22 in-depth interviews with faculty, administrative leaders, and graduates (see Fig. 7.3). The

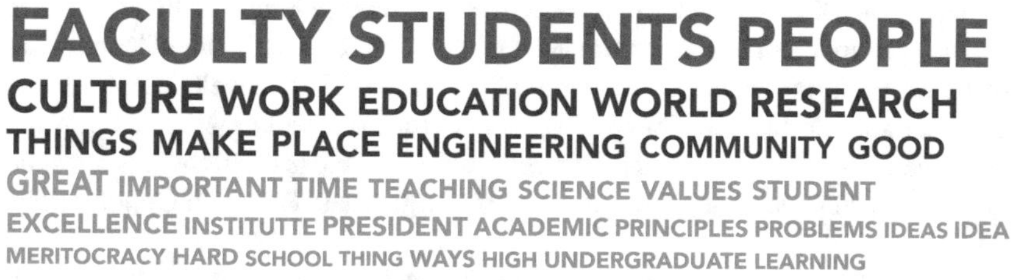

FACULTY STUDENTS PEOPLE
CULTURE WORK EDUCATION WORLD RESEARCH
THINGS MAKE PLACE ENGINEERING COMMUNITY GOOD
GREAT IMPORTANT TIME TEACHING SCIENCE VALUES STUDENT

EXCELLENCE INSTITUTTE PRESIDENT ACADEMIC PRINCIPLES PROBLEMS IDEAS IDEA

MERITOCRACY HARD SCHOOL THING WAYS HIGH UNDERGRADUATE LEARNING

Fig. 7.3 A word cloud was generated automatically, based on word frequency in oral histories and interviews

attributes are presented in no particular order, and without implication that they are all of equal importance:

- Openness (To criticism, To change, To new ideas): Open education, open research, open community
- Excellence (Demanding, Intense, Community): Excellence and accomplishment, excellence and community
- Innovation (Playful, Creative, Entrepreneurial, Irreverent): In the campus and community, creativity and passion, entrepreneurial ecosystem
- Usefulness (Pragmatic, Minds and Hands, Impact): Problem-solving and data driven, interdisciplinary, work with industry, inventions, experiential learning
- Freedom (Trusting): Freedom to pursue ideas and interests, freedom to make mistakes
- Focus (Compact, Collaborative, Unity): One faculty, faculty and administration, breadth and depth, teaching and research)
- Principled (Takes a Stand, Passion, Leadership): Mission and values, leadership

In many ways, these points reflect the design of MIT founder William Barton Rogers, who argued that "in industrial society, science and technology were legitimate foundations for higher knowledge." He wrote that a university has a responsibility "to apply the fruits of scientific discovery to the satisfaction of human wants." These very much are reflected in the characteristics called Usefulness and Principled.

Yet culture can and does evolve—culture that does not adapt to a changing environment will surely die. MIT's culture has evolved in several episodes. It was founded as an institute of technology, but by the early twentieth century, resembled more of a trade school. The decision by the trustees to hire a prominent scientist as president (Karl Taylor Compton) had a strong influence on pre-World War II MIT culture. MIT went from a trade school to major center of science in 30 years. This change emerged from the opportunity to strengthen MIT along the lines of its original principles; the threat of WWII; and the leadership of the trustees and president.

Those familiar with MIT will recognize the underpinnings of this culture in recent developments. For example, the creation of MIT OpenCourseWare matches with the cultural attribute of openness. The development of the Schwarzman College of Computing stems from the willingness to innovate. And, the trend of focus and unity are reflected in the close relationships between faculty and students that grow out of the Undergraduate Research Opportunities Program.

Prepared based on input from the MIT Skoltech Initiative; and Dick K.P. Yue, the Philip J. Solondz Professor of Engineering, Professor of Mechanical and Ocean Engineering, Massachusetts Institute of Technology, Cambridge, Massachusetts, USA.

Reference

1. Keyser SJ (2011) Mens et Mania: the MIT nobody knows. MIT Press, Cambridge, MA

7.3.2   Engaging the Community in a Discussion of Culture

Practical steps can be taken by thought leaders at the university to stimulate the evolution of culture. For example, taking the initiative to define the university's economic impact as rich and complex precludes others from defining it narrowly and threateningly [18]. Involvement of the community leads to impact measures [19] that reflect university culture. Stakeholders will also be excited and motivated by this discourse.

Including among the university's principles or values some that support regional or national economic development is an important enabler for cultural evolution. The debate on these principles can be wide and open. It is particularly important to engage new members of the community, who tend to bring fresh perspectives. If formulated carefully, the principles reflect the existing culture blending with the new. If the principles are slogans only, they are guaranteed to lose hearts and minds.

There are several circumstances that simulate the evolution of culture. External crises certainly do so. There can be a national debate about the role of universities, a financial recession, or a societal calamity. There are externally driven opportunities as well: a new national initiative or funding program, or a request from a major stakeholder for critical help. When leadership changes at any level, a window in time opens in which new visions can be created. Sometimes universities react to creative actions of respected peers, or to wise advice from a respected friend.

Culture sometimes evolves from the bottom-up, stimulated by visible actions or statements by thought leaders at the university. The activities of university programs or units are often not well known across the campus. If culture evolves in a desired direction somewhere, it is useful to make others aware. Celebrating evidence of cultural change is a powerful signal.

7.3.3 Shaping an Aligned Student Subculture

A special and powerful factor in university life is the student subculture. It is largely self-organized and acts autonomously from faculty and administration. In bringing about cultural evolution, the student subculture is likely to be involved and can be supportive or confrontational. Smart leadership engages students upfront in any cultural evolution [20].

Logically, students should be advocates for the broader mission in economic development. The broader mission informs education in skills and provides participation in research and with industry. These are all investments in the students' future. It can also make the university more engaging and fun.

At established universities, encouraging students to take initiatives and soliciting their opinions can be empowering. If things move too slowly, students often want to speed things up. At new programs or universities, explicit steps can be taken to establish student subculture. For example, at Skoltech a group of students was admitted to a "year zero" and sent to major universities in the USA, Europe, and Asia to absorb student innovation culture there. They returned as "founders" to build the student experience at Skoltech.

7.4 Mission and Strategic Planning

7.4.1 Revising the University Strategy Based on a Mission that Includes Innovation

The strategy of a university sets its long-term goals and provides a notional roadmap of how to achieve them. All universities now publish their mission, vision, and goals in the form of strategic plans. The quality of these plans is steadily improving as they assume greater importance. Universities are becoming bolder at benchmarking against others; they claim distinctiveness, state priorities, and measure performance. Mission and strategy can be developed for an entire university, but also usefully for units or programs. In a university, strategy is not intended to prescribe the scholars' activities, but to facilitate them.

Box 7.4 Goals of Mission and Strategic Planning

Universities will more effectively support their academic programs and create a more supportive intellectual ecosystem, all while adapting to their expanded mission in innovation,

By revising the university's mission, strategy, and priorities

Focusing its investment of resources and communicate how the university will contribute and distinguish itself, including in innovation.

Developing strategy involves focusing investment and effort. It is developed through a process that reflects the culture of the university and the needs of its stakeholders. The objectives of strategy are to guide the university in defining its mission (core purpose), its vision (what it hopes to achieve in the long term), and its scope of activity (where it will contribute). Strategy also leads to setting goals and launching initiatives. It establishes priorities that provide important guidance in decision-making. As an instrument of communications, strategy provides a coherent explanation of the aspirations of the university and should help to attract resources [21].

The need to revise mission and strategy stems from the demands faced by contemporary universities. Traditionally, universities relied more on internal inputs and focused most on outcomes in education and research [22]. Now that they act as accelerators of innovation, it makes sense for them to become more adaptable and to widen the range of stakeholders. They must consider new opportunities, build new internal strengths, and revise mission and goals.

7.4.2   Making Strategy through Engagement and Evidence

An approach to strategic planning is sketched in Fig. 7.4. It begins with a recognition of culture, which also will be evolving based on the expanding mission (Evolving Culture). External stakeholders can participate. The ideal representatives

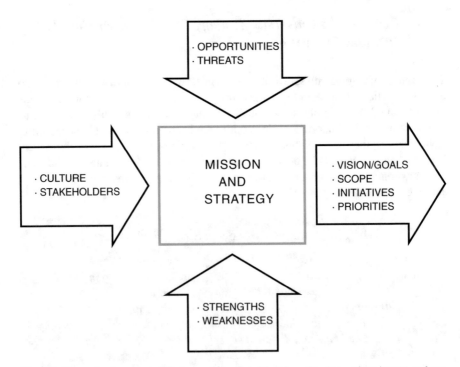

Fig. 7.4 The main elements in Mission and Strategic Planning, showing various inputs, and outputs of vision, goals, etc

are respected thinkers who are knowledgeable about the university (Engaging Stakeholders). The university can decide to take evidence from these external stakeholders or invite them to participate in the process itself.

Making strategy is facilitated by an internal consideration of strengths and weaknesses and an external examination of opportunities and threats [23]. These may have changed because of the new mission. Partners in industry and academia will bring new strengths. Industrial partners, in particular, will have insights about new opportunities.

In an academic setting, constructive discussions are built on engagement and evidence. As the value of strategic planning is as much about process as about the end result [24], engagement of all internal community members is desirable. Engagement will give rise to debate, disagreements, and conflict, but has the best chance of success. It is no secret that many of the best strategic plans are consigned to the bookshelf unless there is ownership at all levels. Making strategy can be the unifying force for the whole community of faculty members, staff, and students. An important outcome of the planning process is that advocates and thought leaders surface.

The collection of evidence is important, and it will often persuade faculty colleagues of these new threats and opportunities. Even top universities now routinely benchmark themselves against peer institutions. They sometimes make broad study visits, and sometimes probe into specific topics. Faculty colleagues are often moved by such peer comparisons. Inputs from thought leaders with specific ideas or proposals can stimulate the discussions. Elements of the history of the university often provide good references. A university should never delegate the task of strategy to a consultancy.

7.4.3 Identifying Priorities, Initiatives, and Collaborations

The outcomes of the strategy process are statements of the high-level mission of a university and its vision and scope. This will have been revised in the light of broadened engagement in economic development. The mission statement, and, particularly, priorities and initiatives, are where the real guidance and insight is presented. These are often stated in terms of desired outcomes in education, research, and catalyzing innovation. The goals also influence how programs are evaluated (Chapter 8).

The priorities shape future decisions. In the modern university, there are simply too many opportunities for a university to follow all. Particularly in science and technology, significant investments are required to develop a scholarly community of critical mass and equip it with state-of-the-art facilities. Universities must prioritize, reflecting both on their own goals and on those of regional or national systems.

An example of effective strategic planning is presented in Case 7.3: ASU—a New American University. Through development and execution of strategy, a previously undifferentiated regional university has become a major research university and a critical element of regional economic development.

Case 7.3 Arizona State University (ASU)—a New American University

A forward-looking strategy, with associated vision, empowerment, resources, and incentives, can be a powerful force in repositioning a university.

In 2002, when I became Arizona State University president, the school was largely still an undifferentiated regional public university. In an era of the increasing power of knowledge, we proposed to reconceptualize ASU as an impactful *and* accessible national research university. I call this model the *New American University* [1], inspired in the writings of Frank Rhodes and James Duderstadt.

The challenge was to provide access to a broad range of students, not just the academically elite, directly addressing the inequity in graduating rates of lower income families. Yet, we wanted to build a faculty dedicated to education *and* discovery.

We reconceptualized ASU as an urban university that combines accessibility to an academic platform underpinned by discovery, inclusiveness to a representative demographic; and, through its breath of functionality, the maximization of impact on society. Key stakeholders, including the Arizona Board of Regents and the local community, supported this new model that assumes responsibility for the economic, social, cultural, and overall health of the community it serves. The model calls for accelerating the clock speed of action. And it stressed that the actions of individuals and the university as a whole be *accountable* to society and the public at large.

We engaged the faculty in creating a culture of aspirational design. We posed the evolution of the university as an ongoing design problem and encouraged design thinking. The process had as elements:

- *Creating a vision:* I used my inaugural to propose a way forward.
- *Building initial momentum:* We ran a campaign to attract early adopters, ideas, and support, resulting in early visible successful demonstrations.
- *Empowering the faculty to be designers:* We granted autonomy and empowered the local team to be in charge, carrying the responsibility of meeting community-based goals.
- *Providing resources and tools:* We supplied tools, support, and project managers to implement the designs at no cost to the local team.
- *Rewarding success*: To enhance faculty thought, we provided new tools, new resources, freedom, and creativity.

Our outcomes are measured against our charter (see Box), mission, and associated goal. Quantitatively, ASU now has three times as many graduates and four times as much sponsored research as in 2002. We have restructured our research to more directly address the grand challenges: for example, the School of Space Exploration is participating in more than a dozen NASA missions and leading the journey to the asteroid Psyche. We have made strong progress in expanding the diversity of students, especially Native Americans. ASU has the highest graduation rate of public universities that provide broad access to students and has lowered the cost to the state for each degree awarded.

> Our experience offers some advice for universities conceptualizing a new future:
>
> - *Find a mission that is unique and build on your strengths and assets in the local context.* Do not accept being one of the crowd.
> - *Focus on excellence of the faculty and access for students.* Universities are distributed human organizations, and their excellence is directly linked to that of the faculty. But, the university also has a responsibility for broad access to students of the region.
> - *Do not accept the status quo.* Universities must constantly evolve. Engage the community in an ongoing process of aspirational design.
>
> Contributed by Michael Crow, President, Arizona State University, Tempe, Arizona, USA.
>
> **Reference**
>
> 1. Crow M, Dabars W (2015) Designing the New American University. Johns Hopkins University Press, Baltimore, MA
>
> **Box in Case 7.3 The ASU Charter**
>
> ---
>
> ASU is a comprehensive public **research university**, measured not by whom it excludes, but rather by **whom it includes** and how they **succeed**; advancing **research and discovery** of public value; and assuming **fundamental responsibility** for the economic, social, cultural, and overall health of the **communities** it serves.
> The Mission is to:
> - Demonstrate **leadership** in academic excellence and accessibility
> - Establish **national standing** in academic quality and impact of colleges and schools in every field
> - Become a **leading global center** for interdisciplinary research, discovery, and development
> - Enhance our **local impact** and social embeddedness
>
> ---

As typified by ASU, universities are no longer islands—all belong to a complex institutional mainland. Strategic plans should include a thoughtful reflection on how the institution fits in the national or regional ecosystem and, consequently, should be bold in stating priorities.

Modern universities require significant capital and resources, and new resources generally are needed to evolve. University planners should interact with principal funding agents. This ecosystem is becoming increasingly complex, with regional, national, and supra-national players that include government, philanthropies, and industry. Often it is necessary to look beyond traditional sources; for example, to programs of regional economic development. In general, a diversity of funding sources permits universities to be flexible.

7.5 Governance

7.5.1 Good Governance in Universities

Governance is the formal way in which a university conducts its activities, in alignment with its mission and culture. Good governance enables the organization to empower all of its people to work together coherently, to use human and financial resources responsibly, and to account for them appropriately. Governance includes policies, structures, and budgets. Good governance provides for clear assignment of responsibility and authority, and for transparent and responsive decision-making.

Box 7.5 Goals of Governance

Universities will more effectively support their academic programs and create a more supportive intellectual ecosystem, all while adapting to their expanded mission in innovation,

By updating decision-making, policies, organizations, and budgets

Promoting a clear definition of devolved authority and transparent decision-making necessary to support the strengthened role in innovation

---

University governance is of interest to government officials who often express skepticism about its effectiveness, parents who are concerned about the cost of education, and the commercial sector concerned about the pace of change. Universities should be ready to explain their governance to stakeholders in light of these concerns.

An eternal tension in university governance is between the academic freedom and prerogatives of the faculty, and the responsibilities of the academic leadership. This is sometimes framed as a dichotomy between top-down ("managerial") and bottom-up ("collegial") organization. To resolve these tensions, we have to ask who has authority, and for what?

We advocate shared governance [25]—where decision-making authority is rationally devolved. Some things are better managed from the bottom-up, including curriculum design and choice of research topics. Some things are better managed from the top-down, including strategic planning and allocation of space. Of course, this is a false dichotomy. Bottom-up initiatives will eventually be accountable upwards. Top-down initiatives will work only if they have support throughout the organization.

Each university develops a system for sharing governance by agreeing where autonomous authority and accountability best resides. This allows the leadership to be effective at leading and risk-taking, and for all members of the community to develop the confidence to find their own solutions. It leaves the administrative leadership in charge of three key decisions: whom to recruit, what incentives to use, and how to allocate financial and space resources.

7.5.2 Updating Governance

Universities need to consider how they will update their governance in view of their expanded economic mission [26]. The principles of good governance will hold: responsibility, accountability, transparency, and responsive decision-making [27]. External stakeholders are likely to be confused by the system of shared and devolved authority [28] but are not likely to interfere in governance if decision-making is clear and effective [29].

Expanding innovation will require some new governance activities, but many will be accommodated in the existing systems. For example, new curriculum leading to innovation will be handled like any educational change. New projects in catalyzing innovation will be handled like any new sponsored research projects. Other changes will require new understandings about governance; for example, university leadership may need to be involved in the creation of large industrial partnerships. In these cases, it is important that this be done transparently, with updates clearly communicated to the community.

These changes will benefit from the involvement of external thought leaders from industry, and small and medium enterprise. The university needs to find a balance between the legitimate need to involve external participants and the sense of independence of scholars [30, 31]. One option is to invite stakeholders to participate in an advisory capacity at all levels.

Involvement in the decision-making process is more complex [32]. While it is common for senior governing bodies to have external members, it is less common further down in the organization. Judicious choice of external members in some key committees could have a beneficial impact.

In some universities, students are involved in academic decision bodies. Students are important stakeholders of the university. They are the owners of the student subculture and the most important agents of knowledge exchange. Involving them in decisional roles has the potential to significantly strengthen the role of the university in society.

The case KTH—Student Influence in University Governance discusses the systematic involvement of student representatives in KTH decision-making bodies, and the long-term benefit it brings to the university and to education (Case 7.4).

Case 7.4 KTH Royal Institute of Technology—Student Influence in University Governance

Viewing students as key stakeholders of their own education can lead to systematic and continuous student influence that will shape the university.

The purpose of the Student Union at KTH (THS) is to improve education. Over decades, student influence has evolved and become increasingly institutionalized in Sweden, especially since 1993 when students gained the legal right to be rep-

Fig. 7.5 Student representative Simon Edström in constructive discussions with President Sigbritt Karlsson (Susanne Hellquist photograph)

resented in all decision-making that affects them. The student representatives to governance bodies are appointed by the student unions [1].

THS consists of program-level chapters and a central university-wide body. The chapters organize students in course and program development, in parallel with their studies. On the central level, two students are appointed to work on improving education full-time, during a 1-year sabbatical. They serve in most organs at KTH, including the university board and all committees and working groups (see Fig. 7.5). Being present at all levels of university governance affords the student representatives a unique vantage point, making the student body a powerful driving force.

A recent example of student influence is in educational development. For many years, THS has demanded that grading criteria must be communicated in the course syllabus, and KTH finally agreed.

This illustrates how the students identified a trend emerging in higher education and pushed for its implementation.

THS also contributes to shaping the future of the university. Student representatives have two seats in every appointment and promotion committee, with full voting rights. Most cases call for discussion and judgment when applying the promotion criteria, including balancing the qualifications for research and education. This is an example of systematic and continuous student influence that can shape the faculty, and, in the longer term, the university. Other examples include program structure development, new internal funding models and reward systems for teaching excellence.

Universities wishing to support organized student influence may consider:

- Including student representatives in boards, committees, and standing meetings.

- Setting up requirements to ensure democratic processes in the student body that appoints representatives.
- Building the competence of student representatives who need to learn quickly, e.g., how the university works.
- Supporting continuity between representatives, e.g., making documentation available to new representatives to learn about the background of issues.
- Not immediately implementing student suggestions, but instead, understanding why the issue is brought up and discussing solutions together, as students may not be aware of all alternatives or understand what could work in the organization.
- Understanding the fluctuating nature of student influence (many student representatives are fantastic but occasional blunders may occur).

Student influence is a cost-effective and democratic way to develop university policy, educational programs, and courses. These experiences are based in the Swedish context, where student influence has been well established for decades. Other cultures may have different norms regarding the roles of students and of educators. The most important aspect is to see students not as customers, but as key stakeholders of their own education. This, in turn, has a cascading effect all the way from individual teacher interactions to top university governance.

Contributed by Simon Edström, former Head of Educational Affairs of the Student Union (THS), KTH Royal Institute of Technology, Stockholm, Sweden.

Reference

1. Student Representatives. In: KTH Royal Institute of Technology. http://ths. kth.se/student-representatives. Accessed 20 Jan 2020

7.5.3 Updating Policy, Budgeting, Organization

Policy provides day-to-day guidance in the execution of functions and decisions. Policy should be clear and firm, but not rigid. Budgeting should be transparent and consistent with strategy and policy [33]. There are several areas of policy that might be updated as a result of the expanded mission. Intellectual property tops the list. Here, universities encounter a tension between the need to protect inventions and the desire to work openly with partners (Chapter 5).

Another important policy is the one that governs faculty involvement in outside professional activities. Universities often allow faculty time for such activities, but also often place limitations on roles that may be played, and on the involvement of student advisees in a company. A closely related issue is the use of campus facilities by industry and start-ups.

Traditional university organizational structures tend to revolve around established disciplines, providing a way of building and transferring disciplinary knowledge. But much new thought is emerging at the boundaries of the disciplines, and problems faced by scholars require integrating approaches from many disciplines.

If the university is to address these issues, then flexible and porous organizations will evolve, reflecting new academic and industry partnerships. Initially, these are likely to be informal collaboration (Chapter 4). In time, a university will likely consider more formal integrative mechanisms, such as interdepartmental centers or schools. In the limit, universities might even choose to do without any disciplinary structures, as has been the case at some newer universities.

7.6 Faculty and Staff Resources and Capabilities

7.6.1 Faculty Members and Professional Staff, the Core Assets

Personnel are the core asset of any organization, and, particularly, of a university. While faculty recruiting is key, it is also essential to assemble a complementary professional workforce that helps execute the university's expanded mission. Once at the university, faculty and staff capabilities should continue to be strengthened through mentoring and professional development.

> **Box 7.6 Goals of Faculty and Staff Resources and Capabilities**
>
> Universities will more effectively support their academic programs
> and create a more supportive intellectual ecosystem, all while adapting to
> their expanded mission in innovation,
> **By recruiting and developing high-quality faculty members and
> professionals**
> *Yielding a university community better able to undertake the tasks and
> programs linked to the strengthened mission in innovation*

While research-intensive universities primarily hire and promote excellent researchers and dedicated teachers, some of these individuals may lack the capabilities or inclination to interact with industry or be agents of knowledge exchange. Once they arrive at the university, many academics receive little formal training to develop these capabilities.

This situation is represented in Table 7.4, where the categories of personnel are shown on the left, and the needed capability on the top. The traditional situation is represented within the dark grey box (upper left). Faculty take the leading role in education and research. Professional research staff and post-doctoral associates support the faculty in research. A variety of teaching ranks, including assistants, lecturers, instructors, and laboratory staff help the faculty in education. These are sometimes supported by a small group of professionals in education: experts in pedagogy, curriculum, and assessment.

Table 7.4 A traditional (dark shade) and expanded (light shade) view of faculty and staff resources and capabilities

	Education	Research	Innovation
Faculty	++	++	+
Research staff		++	
Lecturers and instructors	++		
Professors of practice	+	+	++
Innovators in residence	+		++
Knowledge exchange professionals		++	++

7.6.2   Faculty and Professionals for the Expanded Mission

Professors cannot be experts at everything. They can stretch, but will need help in catalyzing innovation and knowledge exchange [34]. Therefore, the university should adapt and develop a new framework, still populated primarily by faculty, but including professionals in complementary roles (Table 7.4). These additional categories are:

- Professors of the practice—distinguished practitioners who have returned to the university, and who can help interpret the needs of society and the potential for university impact. These contribute primarily to innovation, but also get involved in research and education. These are often longer term and full-time appointments. Good practice is to appoint, fund, empower, and value these professionals as full members of the faculty, and not to treat them as adjunct or lesser members of the community. Professors of the Practice are common in medicine and architecture, and are extremely valuable to collaborative research and innovation.
- Innovators and entrepreneurs in residence. They are often serial entrepreneurs who play a role in catalyzing innovation in the university, particularly in the practice of University-Based Entrepreneurial Venturing (Chapter 5). These professionals will be of enormous help in creating ventures and will contribute to teaching about innovation. Their presence as experts and role models will have great influence on students. They are often short-term or part-time appointments.
- Professionals at the university–industry interface who are experts at knowledge exchange. These include experts in licensing and intellectual property, and in negotiations with companies for funded research. They also include managers of the interface with industry, who make introductions between industry and enterprise and faculty, and assist in Facilitating Dialog and Agreements (Chapter 5).

All of these professionals deserve respect and standing at the university. They must be appropriately empowered to execute their responsibilities.

7.6.3   *Recruiting, Hiring, and Career Development*

If personnel are the core assets, the gateway process is the recruiting and hiring of strong potential contributors. In the global academic network, there is strong competition for academic talent. The university should develop a strategy to recruit capable individuals in all of the categories of Table 7.4, but particularly the faculty. The mission in economic development will, on the one hand, make faculty recruiting more difficult, because of the additional roles that faculty are asked to play. But a compensating factor is the attractiveness to some scholars of working in an environment that values the impact of their work. Faculty and staff recruiting should also respect the local norms for diversity.

Case 7.5 Technical University of Denmark (DTU)—Development of Faculty Teaching Competence

New and experienced faculty members participate in structured and comprehensive teaching development programs that build capacity to design courses as effective environments for student learning.

To support new faculty members in their role as engineering educators, DTU has since 2004 offered a training program called *Education in University Teaching at DTU* (UDTU). It is part of a larger introduction scheme, which also includes collegial mentoring and supervision of teaching during the introduction period, and development of a teaching portfolio. Successful completion is required for appointment as an associate professor, and for taking on full course responsibility.

Each year, about 70 participants start UDTU, which consists of four modules with a total estimated workload of 250 h (see Fig. 7.6). The focus is first on classroom management, followed by topics relevant for designing, implementing, and evaluating courses, and for understanding teaching and assessment from a student learning perspective. The final module is a project in which participants develop their own teaching, using action research methods to analyze the student learning outcomes and the student course experiences. During the project, the participants submit two reports and continually get feedback on their work.

The main objective of UDTU is to develop the capacity to design courses as effective environments for student learning. It aims to develop participants' criti-

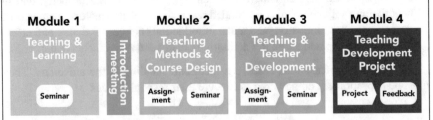

Fig. 7.6 The training program called *Education in University Teaching at DTU*

cal and creative sense for their own teaching practice [1]. The learning paradigm is constructivism, with a theoretical framework based on constructive alignment [2] and approaches to learning [3]. The theory is closely related to teaching practices at DTU, emphasizing participants' application in their own teaching.

Following the success of UDTU, departments have requested equivalent activities for experienced teachers. Consequently, *University Pedagogy for Experienced Teachers* (UP) has been held three times [4]. The estimated UP workload is 140 h over one semester. The setup is similar to UDTU, but customized for experienced faculty, based on Kolb's model of experiential learning [5]. The starting point is to explore and analyze the participants' own teaching practices from a student perspective (i.e., based on interviews with students). This is followed by theory on teaching and learning, course planning, orchestration of learning, assessment, and research-based teaching. Each participant presents an implementation plan for revising a course, with feedback and coaching from peers and course leaders, and, finally, writes a teaching statement.

To ensure joint ownership, the activities are coordinated by the teaching development unit, LearningLab DTU, and carried out in close cooperation with the departments. Both UDTU and UP are taught by teams of two course leaders, one faculty member from a DTU department and one LearningLab DTU staff. In addition, previous participants visit UDTU to share how they keep developing their teaching practice inspired by the ideas from the program. In UP, the participants also meet leading Danish teaching and learning experts. There are indications that the program is working as intended. Most importantly, the departments find it useful and stay involved and supportive.

Contributed by Pernille Andersson, Educational Consultant, LearningLab, DTU; Nina Qvistgaard, Senior Executive Educations Officer, National Institute of Aquatic Resources, DTU; Claus Thorp Hansen, Associate Professor, Department of Mechanical Engineering, DTU, Kongens Lyngby, Denmark.

References

1. Andersson PH, Onarheim B (2015) Facilitating creativity as a core competence in engineering education. In: Proceedings of the 43nd Annual SEFI conference. Orleans, France
2. Biggs J, Tang C (2011) Teaching for quality learning at university. Open University Press
3. Marton F, Säljö R (1984) Approaches to learning. In: Marton F, Hounsell D, Entwistle NJ (eds) The experience of learning. Scottish Academic Press, Edinburgh, pp. 36–55
4. Andersson PH, Hansen CT (2014) Experiential learning as a method to enhance experiential teaching. In: Proceedings of the 42nd Annual SEFI Conference. Birmingham, UK
5. Kolb DA (2014) Experiential learning: experience as the source of learning and development, 2nd edn. Pearson Education, Upper Saddle River, NJ

Provision should be made for career development for all categories of personnel, including the faculty. Opportunities center on internal development programs, mentoring and inclusion in communities of practice within the university. New professors are usually paired with a faculty research mentor in research and might also be paired with an innovation mentor—perhaps a professor of the practice.

An early career development program introduces new faculty members to their roles in education. Most new faculty are hired based on demonstrated research capability. These faculty need assistance to become proficient teachers and master new educational schemes. In some systems, a teaching credential is required before an early career faculty member can lead a course by themselves. Case 7.5 DTU—the Development of Faculty Teaching Competence elaborates one comprehensive approach to teacher development in which new faculty learn how to make courses effective environments for student learning.

A wider view of faculty education development is presented in the case KTH—Faculty Development Intertwined with Education, Organization, Students, and Research. In Case 7.6, we learn about a comprehensive program for faculty development, which still supports individuals and their course development. But it is intertwined with activities to develop the educational programs and organization of the university.

Mid-career development cultivates faculty and other professionals for senior positions of leadership and responsibility at the university and beyond. This can take the form of mentoring by more senior members of the community, and professional exchanges with industry, government, and other academic laboratories. Good practice includes a reflective personal plan with goals and activities tailored for individual growth.

Case 7.6 KTH Royal Institute of Technology—Faculty Development Intertwined with Education, Organization, Students, and Research

Enabling and driving the development of education through aligned activities in teaching, programs, and organization.

In the past, KTH faculty development focused on individual educators developing their own teaching. Today, there is a comprehensive program for faculty development that still supports individuals and their course development, but it is intertwined with activities to develop the educational programs and organization. Together this constitutes a strategy to *enable and drive* the development of education through well-aligned and long-term persistent activities.

On the level of individual faculty development, KTH seeks to balance its demands on faculty with support. On the support side is a comprehensive program customized for KTH faculty (see Fig. 7.7). The demand is defined by the academic career system, which, since 2003, requires faculty on all levels to have at least 10 weeks of formal education on teaching and learning in higher education. This requirement has been reaffirmed on the national level by a commission

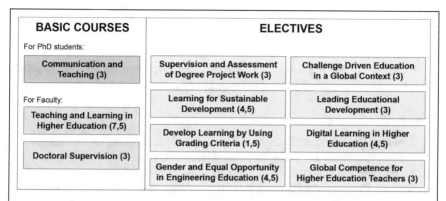

BASIC COURSES	ELECTIVES	
For PhD students:		
Communication and Teaching (3)	Supervision and Assessment of Degree Project Work (3)	Challenge Driven Education in a Global Context (3)
For Faculty:	Learning for Sustainable Development (4,5)	Leading Educational Development (3)
Teaching and Learning in Higher Education (7,5)	Develop Learning by Using Grading Criteria (1,5)	Digital Learning in Higher Education (4,5)
Doctoral Supervision (3)	Gender and Equal Opportunity in Engineering Education (4,5)	Global Competence for Higher Education Teachers (3)

Fig. 7.7 Courses on teaching and learning customized for KTH faculty, with number of ECTS credits (1.5 credits corresponding to 1-week workload)

led by the KTH president, Sigbritt Karlsson. Candidates for hiring and promotion must also demonstrate teaching competence, documented in a teaching portfolio. The meaning of teaching competence has developed over the years, from focusing mainly on individual classroom performance, to the whole teaching mission, including making contributions of a collegial, institutional, and scholarly nature.

The educational development emphasizes the program and organizational levels. KTH was one of four co-founders of the CDIO initiative in 2000 [1]. The aim is to meaningfully integrate the development of theoretical understanding and professional skills throughout an educational program, while strengthening connections between courses to ensure progression. This systematic program-driven approach has become the norm for developing KTH programs and courses.

To further identify needs and opportunities, KTH has initiated evaluations involving international panels [2]. As a result, the remit and mandate of various educational leadership functions were clarified. Further, KTH-wide forums for networking and dialog have been created to stimulate reflection and deliberation, and to speed up action for dealing with small and large issues. One network gathers departmental directors of study; another, the directors of all 120 educational programs. There is also a general meeting open for all and working groups on various prioritized issues. Student representatives exert considerable influence as they are deeply involved in all these activities and have seats in every decision body.

To support these developments, KTH has a long-standing educational development unit. In dialog with university leaders, it provides the faculty development program and coordinates many of the activities mentioned above. It also organizes week-long workshops for visiting groups of international faculty. Since 2011, the unit has evolved into a center for engineering education research with a PhD program. In 2018, it was a key part in forming the KTH Department of Learning in Engineering Sciences, with several professorships. Research studies are often made in collaboration with educators and educational leaders throughout KTH. The aim is to integrate educational development and research.

Contributed by Anna-Karin Högfeldt, Lecturer, Director of the Faculty Development Program, KTH Royal Institute of Technology, Stockholm, Sweden.

References

1. Crawley E, Malmqvist J, Östlund S et al. (2014) Rethinking engineering education: the CDIO approach, 2nd edn. Springer, Cham
2. Karlsson S, Fogelberg K, Kettis A, Lindgren S (2014) Not just another evaluation: a Comparative Study of Four Educational Quality Projects at Swedish Universities. Tert Educ Manage 20:239–251

7.7   Academic Facilities

7.7.1   Facilities that Enable and Adapt to the Expanded Role in Innovation

Facilities are the university's physical and informational backbone of a university. They can be icons of considerable tradition and power, like the Long Room at Trinity College Dublin, the old gate at Tsinghua, and the Stalin gothic tower of Moscow State University. Functionally, facilities include classrooms, social spaces, laboratories, project workshops, and digital networks, all of which traditionally support education and research.

Facilities at universities with an expanding role in innovation must adapt and support activities associated with catalyzing innovation [35], such as educational programs in innovation [36], new start-up ventures and projects to mature technologies. Knowledge exchange with industry might require facilities for networking and digital collaboration [37].

Box 7.7 Goals of Academic Facilities

Universities will more effectively support their academic programs and create a more supportive intellectual ecosystem, all while adapting to their expanded mission in innovation,

By providing functional, flexible, and available facilities—classroom, social spaces, laboratories, workshops, and IT connectivity

Enabling new activities in learning, innovation, and collaborative research with industry

Good facilities can be a competitive advantage in research and innovation [38]. Facilities are important in attracting new faculty and students. Scholars are more likely to join a university if they have access to laboratories that support their research. Tours of campus for prospective students often emphasize facilities.

Spaces can inspire and motivate and can celebrate historical and contemporary accomplishments. They can help build a sense of ownership and community. Spaces should be well designed and can introduce elements of art, design, and culture. They should provide choice, so that users have a sense of variety in the type of space available for use. There should be public and private spaces, so that some can collaborate while others concentrate in private.

Facilities should be designed to be adaptable, so that they can respond to changes in use patterns. Long-term adaptability of space is desirable and short-term re-tasking of existing space is common. Project workshops, maker spaces, and laboratories must be safe. They should also be available to faculty and students during the long hours these individuals require access.

New facilities can signal new undertakings. If you want to attract innovators, provide new maker spaces. If you want to support community, provide new networking and social space. The types of spaces needed at a university include: class, lecture, and conference spaces; project and maker spaces; laboratory spaces; and spaces for social learning. Education, research, and innovation can occur in all these spaces.

7.7.2 Facilities for Education and Community

Learning spaces are central to the educational mission and should provide for all modes of education, including conventional, digital, experiential, and social. Consideration of the evolving practices in education will affect the traditional mix of amphitheaters, tiered classrooms, flat classrooms, seminar rooms, and project spaces.

Project spaces play an important part in project-based learning. Students are drawn to these spaces by both required and self-initiated projects. Some of these are projects about discovery. Some are about making and innovating. In both cases, students are more likely to be effective in the project if there is some nearby space to promote social learning.

Facilities must increasingly support digital learning, which means digital design studios and rooms for digitally supported education and collaboration. High reliability network connectivity and access to significant computational capability are vital.

A consideration in space planning is the creation of social spaces to facilitate collaboration and improve the quality of life. People at a university spend a lot of time in discussion with colleagues, students, sponsors, and mentors. Social spaces support networking among the entrepreneurial community. These and other interactions are supported by desirable and comfortable space.

In the case MIT—CDIO Project Learning Workshop/Laboratory, new, functional, and attractive spaces were created for research and innovation projects and social learning (Case 7.7). The very layout of the space reinforced the educational paradigm that engineering is about conceiving, designing, implementing, and operating.

Case 7.7 The Massachusetts Institute of Technology (MIT)—CDIO Project Learning Workshop/Laboratory

Not only can teaching laboratories support project, experimental and community learning; their design can signal a shift in pedagogic approach.

Working with international partners, the MIT's Department of Aeronautics and Astronautics created a model for engineering education set in the context of *conceiving, designing, implementing, and operating* (CDIO) systems and products [1]. To facilitate this educational reform, the department undertook a study of how engineering education practices could be improved to better enhance learning.

One area of examination was the more systematic inclusion of project-based learning throughout the undergraduate education. This required the development of associated project learning spaces. An early result was the identification of a variety of learning modes used by students in such spaces. Traditionally, labs were filled with experiments that reinforced disciplinary knowledge, but were only used a handful of times each year. This was viewed as an underutilization of an important and expensive resource.

By observing the student use of such space, we noted students working in three other modes that we call knowledge discovery, system building, and community building. Knowledge building is largely done through experiments: the collection of data and analysis. This is how scientists work. Engineers are engaged in repeated authentic cycles of designing, building, and testing—what we call systems building. We observed that if you create desirable spaces and fill them with appropriate tools, the students will take ownership, come more often, and fill the space with emerging community.

As we started to develop concepts for the space, we identified a set of desirable themes. We determined that the environment must be flexible and easily reconfigured, accessible whenever needed, and scalable to accommodate projects of all physical sizes. It should be sustainable within a normal operating budget, "wired" for power and data access, and coordinated with other major MIT facilities.

To reinforce the concept that the role of engineers is to conceive, design, implement, and operate systems, we decided to use this as the spaces' organizing principle. The branding and organization of each space would signal the new pedagogy (see Fig. 7.8).

Conceive spaces would allow students to envision new systems, understand user needs, and develop concepts. These spaces would emphasize reflection and reinforce human interaction. They would have sufficient technology for communications and information retrieval, but not for design or computation.

Design spaces would support the new paradigm of cooperative, digitally supported design. They would allow students to design, share designs, and understand interaction. They would include a central room for large group interaction and be connected to breakout rooms for smaller teams to work on their projects. They would be in proximity to build space, reinforcing the design-build connection.

Implement spaces would allow students to build small, medium, and large systems. They would offer mechanical, electronic and specialty fabrication, and

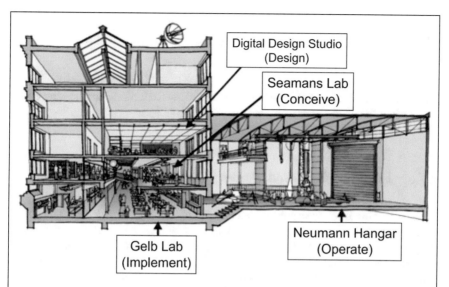

Fig. 7.8 Branding and organization of the spaces reflects the pedagogy

Fig. 7.9 The Hangar—the operate space

software engineering. A key element (and challenge) would be to make them safe, yet as accessible as possible outside "regular" school hours.

Operate spaces would create opportunities for students to learn about engineering operations. There, they could conduct their experiments and projects and simulate operations of real systems (see Fig. 7.9).

> Prepared based in input provided by Cory Hallum, MIT Laboratory Project Manager, and Steven Imrich, Fellow of the American Institute of Architects.
>
> **Reference**
>
> 1. Crawley E, Malmqvist J, Östlund S et al (2014) Rethinking engineering education: the CDIO approach, 2nd edn. Springer, Cham

7.7.3   Facilities for Research and Catalyzing Innovation

Research laboratories are a specialized part of university facility design. They often contain sophisticated equipment. Universities encouraging impactful research will do larger scale and collaborative projects and will have special laboratory needs.

Research facilities are generally categorized as IT, dry, wet, and specialized research spaces. IT requires only rooms and connectivity. Dry labs support physical, mechanical, and electrical work. Wet labs deal with chemically active, toxic, or biological materials. Specialized facilities include wind tunnels, nuclear reactors, animal care facilities, and clean rooms. Benchmarking of these kinds of space at other universities gives insight into needs and costs.

A research space planning process for these costly and complex assets should ensure that existing facilities are up to date and well used. This can be challenging in conventional labs that sometimes support small groups and even single researchers.

An alternative organization is proposed in Case 7.8 called **KAUST—One Lab**. Eleven different core labs are operated by a centralized "one lab" organization. This arrangement provides high utilization, effective service, and well-maintained modern equipment.

> **Case 7.8 King Abdullah University of Science and Technology (KAUST)—One Lab**
>
> *The central governance and operation of research facilities provides enhanced support for researchers and improves efficiency.*
>
> In contrast to traditional models where university laboratories are run by individual groups, the Core Labs at KAUST in Saudi Arabia, are run as a centrally governed "One Lab" facility. By grouping the services of 11 specialized labs, the One Lab model provides better experimental support, while enhancing overall efficiency.
>
> The fully equipped labs are located in seven separate campus buildings totaling an area of 570,000 square feet. The lab's 350 million USD capital base supports 11 scientific fields and research areas: Analytical Chemistry, Animal

Fig. 7.10   KAUST Research Vessel operated and managed by Coastal and Marine Resources Core Lab

Fig. 7.11   Visualization core lab

Resources, Bioscience, Coastal and Marine Resources, Imaging and Characterization, Nanofabrication, Supercomputing, Visualization, Plant Growth, Prototyping and Fabrications, and Radiation Labeling (see Figs. 7.10 and 7.11). The labs house more than 800 instruments and employ more than 200 staff, most of whom have advanced degrees in science and engineering [1].

The One Lab offers more than 870 services, trains more than 2500 people per year, and conducts more than 5000 transactions per month. Transactions are defined as the four types of services conducted in the labs: full service, where lab technicians and scientists undertake work as requested by a user; training, where staff train users on specified instruments; independent use, where the user conducts an experiment in a lab utilizing the equipment after passing the necessary training; and collaboration, where staff work with internal and external users, focusing primarily on new method development.

The One Lab organization brings benefits to the researchers. The equipment becomes a hub of researchers using the same experimental techniques. The equipment is available 98.5% of the time and is always operating within calibration standards. There are expert operators and training available to investigators. The equipment is state-of-the-art and because of the close relationship between KAUST and vendors there is feedback to manufacturers that sometimes leads to upgrades that can preferentially benefit KAUST investigators. The Core Labs receive more than 95% positive feedback from users and has contributed to KAUST attaining the most citations per faculty in QS World University Rankings.

The Core Labs bring multiple organizational benefits, such as finance and resource planning, asset management, user engagement and services, staff recruitment, and utilization policy. In addition, One Lab maximizes equipment availability, reduces equipment duplication, minimizes the cost of maintenance, decreases safety incidents, and simplifies the process by which users obtain the space and equipment they need. Core Lab efficiencies and close relationships with vendors consistent with larger contracts and professional management have led to significant capital and operating savings.

The One Lab model enables KAUST to engage companies and other users with diverse requirements in a straightforward and professional manner. It offers a broad menu of services, all in one place. Currently, KAUST's Core Labs service 1500 KAUST internal users, 21 other universities, and 11 government organizations within Saudi Arabia.

Contributed by: Justin Lee Mynar, Associate Vice President for Research and Executive Director, Core Labs, King Abdullah University of Science and Technology, Thuwal, Saudi Arabia.

Reference

1. King Abdullah University of Science and Technology (KAUST)—One Lab. https://corelabs.kaust.edu.sa. Accessed 20 Jan 2020

Catalyzing innovation also uses space. Student innovators need maker spaces for projects, and these spaces, too, can be IT, dry or wet labs. There is often shared space on campus for start-ups. Good practice is to have suppliers and companies nearby so that they can help innovators build prototypes.

IT entrepreneurs largely have no capital expenses, but non-IT innovators do. They will be more competitive if the university provides capital equipment and research space at low cost or as part of an equity investment. Each university must make its own policy on the use of research labs and equipment in support of start-ups.

7.8 Summary

In this chapter, we present six practices that support the academic practices of the university in education, research, and catalyzing innovation. As presented, they emphasize the adaptive nature of the university that will allow it to transition to an expanded role in knowledge exchange, innovation, and, ultimately, economic development. The supporting practices are:

Engaging Stakeholders, in which the university identifies the broader set of stakeholders associated with its expanded mission, engages with them, and identifies their needs. Since the university is also a stakeholder, its needs are identified as well. The outcomes of this practice influence Mission and Strategic Planning, and Governance. They may help to shape the *Evolving Culture*.

Evolving Culture uses this university transformation as an opportunity to reflect on the values, beliefs, habits, and collective knowledge of the university community. It enables a gradual shift in beliefs, so that the mission of supporting economic development becomes an accepted and valued aspect of culture. The outcomes of this practice have pervasive influence, especially in *Mission and Strategic Planning, and Governance*.

Mission and Strategic Planning calls for a revision of the university's mission, vision, goals, and action programs to support innovation. It provides prioritization and focus for programs, and it proposes decision guidance for faculty and administration. The outcomes of strategy are priorities and actions that have significant influence on *Governance* and *Faculty and Staff Resources and Capabilities*.

Governance is the formal way that authority and responsibility are devolved in the university and is reflected in responsive decision-making, clear policies, porous organizations, and transparent budgets. It has strong influence on the operations of the university, especially in *Faculty and Staff Resources* and *Facilities*.

Faculty and Staff Resources and Capabilities deals with the core assets of the university—people. Both the faculty and professional staff must be convinced to join the university and should be presented opportunities for professional developed, especially in innovation. The outcome is the team which will be the agent of the academic programs.

Academic Facilities, both physical and informational, are the backbone of the university. They span classrooms and social spaces, laboratories and workshops, and IT networks. They enable new ways of learning, research and innovation, and collaboration. The outcome is the infrastructure that will be the instrument of the academic programs.

The careful reader will note that we have introduced eleven academic and six supporting practices, and none is explicitly called "students." Where are the students? They are *everywhere*! The educational practices ensure student learning. The students participate in research, maturing technologies, innovation, and entrepreneurship. The students are stakeholders, shape culture, and participate in governance. They are, in some sense, the real central actors of the university, and they are the key participants in knowledge exchange.

Not surprisingly, the six supporting practices have strong connections to other frameworks on which we have built. Van Maanen describes three perspectives through which one can view an organization: culture, strategy, and governance. The CDIO community has standards that broadly encourage stakeholder engagement, development of faculty and staff, and availability of modern facilities.

A more conceptual linkage is to the work of Burton Clark, especially as interpreted by Van Vught. Clark's article is about "universities that wish and try to adapt to changing environmental conditions." It builds on his five basic characteristics of successful innovative universities:

- "*A* strong steering core" that is quicker, more flexible, and more focused in reacting to demands, much like our practice of Governance.
- "A development periphery." These are mechanisms that relate to the outside world and reach across traditional boundaries. This touches on the practices of Engaging Stakeholders.
- "A diversified funding base," obtaining a widened set of resources and particularly discretionary funds, and lessening dependent on government. We mentioned this in Mission.
- "A strong academic heartland," academic units that accept an entrepreneurial role, which is implicit in the four educational practices.
- "An integrated entrepreneurial culture" that embraces change, discussed in Evolving Culture.

References

1. Van Maanen J (2017) Organizational change: three perspectives from John Van Maanen. In: MIT Sloan Executive Education. https://executive.mit.edu/media-video/mit-sloan-executive-education/three-perspectives-on-organizational-change-with-john-van-maanen/fdfgae9-lpy. Accessed 20 Nov 2019
2. Ancona DL, Kochan TA, Scully M, Van Maanen J, Westney DE (2004) Managing for the future: organizational behavior and processes, 3rd edn. Cengage Learning
3. Klein E, Woodell J (2015) Higher education engagement in economic development: foundations for strategy and practice. Association of Public & Land-Grant Universities, Washington, DC
4. Ryan JH, Heim AA (1997) Promoting economic development through university and industry partnerships. New Directions for Higher Education. https://doi.org/10.1002/he.9704
5. Jongbloed B, Enders J, Salermo C (2008) Higher education and its communities: interconnections, interdependencies and a research agenda. High Educ 56:303–324
6. The Organisation for Economic Co-operation and Development (2000) The response of higher education institutions to regional needs. https://doi.org/10.1787/9789264180550-en. Accessed 4 Nov 2019
7. Lester R (2005) Universities, innovation, and the competitiveness of local economies: local innovation systems project. MIT, Cambridge, MA
8. Coleman MS (2018) Recommitting to America's Unique Government-University Partnership. Association of American Universities. Change 50:128–132
9. Sandmann LR, Plater WM (2009) Leading the engaged institution. N Dir High Educ 2009:13–24

10. Burrows J (1999) Going beyond labels: a framework for profiling institutional stakeholders. Contemp Educ 70:5

11. Kotter JP, Heskett JL (1992) Corporate culture and performance. Maxwell Macmillan International, New York, NY

12. Schein EH (2010) Organizational culture and leadership, 4th edn. Jossey-Bass, San Francisco, CA

13. Hofstede G, Hofstede GJ, Minkov M (2010) Cultures and organizations, 3rd edn. McGraw-Hill, New York

14. Geertz C (1973) The interpretation of cultures. Basic Books, New York

15. Schein EH, Schein PH (2019) The corporate culture survival guide, 3rd edn. Wiley, Hoboken, NJ

16. Deal TE, Kennedy AA (1982) Corporate cultures. Addison-Wesley, Reading, MA

17. Denison DR (1997) Corporate culture and organizational effectiveness, 2nd edn. Denison Consulting

18. Franklin NE (2009) The need is now: university engagement in regional economic development. J High Educ Outreach Engagem 13:51–73

19. Association of Public & Land-Grant Universities (2014) New metrics field guide measuring university contributions to the economy. Association of Public & Land-Grant Universities, Washington, DC

20. Coates H, McCormick AC (2014) Engaging university students: international insights from system-wide studies. Springer, Singapore

21. Morrill RL (2007) Strategic leadership: integrating strategy and leadership in colleges and universities, ACE/Praege. Rowman, Littlefield Publishers & American Council on Education

22. Hardy C, Langley A, Mitzberg H, Rose J (1983) Strategy formation in the university setting. Rev High Educ 6:407

23. Porter ME (1998) Competitive strategy: techniques for analyzing industries and competitors. Free Press, New York, NY

24. Dooris MJ, Kelley JM, Trainer JF (2004) Strategic planning in higher education. New Dir Inst Res 2004:5–11

25. Olson GA (2009) Exactly what is "shared governance"? The Chronicle of Higher Education

26. Fielden J (2008) Global trends in university governance by The World Bank. Washington, DC

27. Edwards M (2000) University governance: a mapping and some issues. Australian Institute of Tertiary Education Administrators. https://www.atem.org.au/eknowledge-repository/command/download_file/id/24/filename/Univeristy_Governance_Meredith_Edwards_2000.pdf. Accessed 2 Jan 2020

28. Amaral A, Mahalhaes A (2002) The emergent role of external stakeholders in European higher education governance. In: Amaral A, Jones GA, Karseth B (eds) Governing higher education: National Perspectives on institutional governance. Springer Science+Business Media, London, pp 1–21

29. Association of Governing Boards of Universities and Colleges (2014) Consequential boards adding value where it matters most: Report of the National Commission on College and University Board Governance. Washington, DC

30. Association of Governing Boards of Universities and Colleges (2010) Policies, practices, and composition of governing boards of independent colleges and universities by The Association of Governing Boards of Universities and Colleges. Washington, DC

31. Association of Governing Boards of Universities and Colleges (2013) Effective governing boards: a guide for members of governing boards of public colleges, universities, and systems by Association of Governing Boards of Universities and Colleges. Washington, DC

32. Larsen IM, Maassen P, Stensaker B (2009) Four basic dilemmas in university governance reform. High Educ Manage Policy OECD 21:33–50

33. Goldstein L (2005) College and university budgeting: an introduction for faculty and academic administrators, 3rd edn. National Association of College and University Business Officers, Washington, DC

34. Etzkowitz H (1998) The norms of entrepreneurial science: cognitive effects of the new university–industry linkages. Res Policy 27:823–833
35. Youtie J, Shapira P (2008) Building an innovation hub: a case study of the transformation of university roles in regional technological and economic development. Res Policy 37:1188–1204
36. Boys J (2010) Towards creative learning spaces re-thinking the architecture of post-compulsory education. Routledge, London
37. Toker U, Gray DO (2008) Innovation spaces: workspace planning and innovation in U.S. University Research Centers. Res Policy 37:309–329
38. Moultrie J, Nilsson M, Dissel M, Haner U-E, Janssen S, Van der Lugt R (2007) Innovation spaces: towards a framework for understanding the role of the physical environment in innovation. Creat Innov Manage 16:53–65

Chapter 8
Evaluation and Expectations at the Adaptable University

8.1 Introduction

8.1.1 Evaluation and Action Linked to Economic Development

As universities grow and become central to the development of regions and nations, stakeholders naturally expect them to document and explain their impact [1]. How does a university know how successful its actions are over time? How can it improve its impact in the future?

Universities do have profound impact on society across a broad spectrum of issues [2–4]. In this book, we highlight university activities in science, technology, and entrepreneurship that contribute to knowledge exchange, innovation and, eventually, economic development. This does not in any way diminish the broader societal contributions of comprehensive and specialized institutions of higher learning.

In this chapter, we discuss how evaluating programs and units helps a university adapt to this expanded mission in economic development. The central task of this evaluation is to validate the university's progress towards effective knowledge exchange in both quantitative and qualitative terms. The university can use that information to continuously improve. We also discuss the practice of establishing expectations for the faculty, and how this practice can support university ambitions.

© Springer Nature Switzerland AG 2020
E. Crawley et al., *Universities as Engines of Economic Development*,
https://doi.org/10.1007/978-3-030-47549-9_8

Box 8.1 Objectives of Evaluation and Expectations

The general objective of evaluation is to provide guidance for improvement of the university's academic activities in education, research, catalyzing innovation and knowledge exchange, and to provide a basis for recognition of success. Establishing expectations for the faculty is part of this broader approach to improvement.

The practices in evaluation and expectation help to mobilize the community and support organizational learning. They provide a way to engage in national evaluation schemes, communicate accomplishment and convey institutional value to stakeholders.

Viewed through the lens of economic development, the specific objective is to more effectively evaluate and report accomplishment, particularly in knowledge exchange. This allows the university to take actions to improve that are consistent with the expanded mission in innovation. Compared to traditional approaches, which evaluate education and research, the scope and timeframe of this new evaluation process is expanded to include the innovation outcomes and the process of knowledge exchange.

8.1.2 Evaluation Based on Community Goals

Any scheme for evaluation starts with a clear understanding of community goals [5, 6]. At this juncture, we assume that the goals of a university have been set, responding to the aspirations of the academic community and the needs of the external stakeholders. The goals will reflect outcomes in education, research, catalyzing innovation and knowledge exchange (Chapters 3–6). Viewed through our lens, these outcomes include talented students, discoveries, and innovation creations that are exchanged at the porous boundaries of the university; these contribute to economic development and the larger social good. The programs to reach these goals have been set out by strategy, are within a system of shared governance, and are reinforced by an evolving culture (Chapter 7).

Robust recurrent evaluations of the university's programs within the context of these goals is no longer an option; they are a necessity. There are several important reasons to evaluate an institution's success:

- To determine whether and where the university is succeeding.
- To inform discussion of how the university can improve.
- To meet the requirements of national evaluation and accreditation schemes.
- To build confidence and support among stakeholders.

The first two bullets are the most pure and altruistic—all organizations should be engaged in continuous improvement, using effective quality processes supported by evidence. The third is a necessity of national quality systems. Such systems aim to maintain minimum standards, but should provide flexibility for universities to adapt the standards to their local situation. The importance of the last reason is growing rapidly. Governments across the world are increasingly calling on universities to provide information on their impact on teaching, research, and catalyzing innovation [7]. And when done well, evaluation can also be a strong force for change [8].

8.1.3 The Practices of Evaluation and Faculty Expectations

There is a wide range of experience in evaluation at universities [9]. Some universities have little experience, while others are expert [1]. We recommend two related practices, as outlined in Table 8.1.

The first practice is the formal periodic evaluation of programs or units. Evaluation judges the overall effectiveness of a program based on evidence of progress towards attaining its goals [10]. Evaluation will have a stronger influence when it systematically triggers actions to improve.

A second practice involves establishing commonly held and broadly supported expectations for the faculty members. These expectations should be relatively general within an academic domain, and stable from year to year. They should be aligned with the norms and mission of the university. This practice takes into account the faculty's unique responsibility to contribute to the university, while being independent. Professors must contribute to the community, and be ready to pivot quickly and explore new ideas, guided by curiosity and their own interpretation of needs. Successful faculty can be recognized, reinforcing the improvements underway.

A virtuous cycle of planning, actions, evaluation, and improvement can be created. Goals and plans are based on community and stakeholder needs; strategy, programs, and faculty expectation influence outcomes; outcomes are documented and evaluated; and evaluation informs the revision of strategy and programs.

The changes that might be brought about by adopting these practices are summarized in Table 8.2.

Table 8.1 The practices of evaluation and expectations

Practice name	Description of the practice > and its outcome
Program Evaluation	Collecting evidence that reflects the university goals and evaluating success of programs or units > Demonstrating the effectiveness and contributions of the university and inspiring action to improve.
Faculty Expectations and Recognition	Establishing expectations and recognizing accomplishments of individuals in education, research, innovation, and knowledge exchange > Better aligning the actions of the faculty with the university goals.

Table 8.2 Reference and aspirational situation for evaluation and expectations

Reference situation	Aspirational situation
Program evaluation is ad hoc and often emphasizes research and reputation	Program evaluation is systematic and supports the missions in education, research, and innovation, leading to follow-through action [Program Evaluation]
Faculty expectations set by tradition and practice, with recognition primarily focused on research	Faculty expectations established by consensus and aligned with university norms and goals, with recognition given across education, research, and innovation [Faculty Expectations and Recognition]

8.2   Program Evaluation

8.2.1   The Benefits and Evolving Challenges of Evaluation

A university has a responsibility to its stakeholders to be thoughtful and earnest in its self-evaluation [11]. The university should base evaluation of programs and units on pre-agreed criteria, documentation of outcomes, and thoughtful reflection on the results. Such a process of evaluation should not drive the university, but it should, instead, reflect the ambitions of the university and help satisfy them.

The types of evaluation, including process and evidence, should be selected by the university, and can reflect its acceptance of an expanded mission in economic development. The results of the evaluation should become known, and they should be acted upon. Such an evaluation should be validated by periodic independent review by external peers and key stakeholders, or by rigorous benchmarking of aspirational peers.

Box 8.2 Goals of Program Evaluation

Universities will improve their academic programs and communicate accomplishments and institutional value to stakeholders, all while adapting to their expanded mission in innovation
By collecting evidence that reflects the university goals, and evaluating success of units or programs
Demonstrating the effectiveness and contributions of the university and inspiring action to improve.

The process can vary in scope. It can serve to evaluate a single program or unit, or the entire university [12]. For small and focused universities, one system may be appropriate for the entire university. For a larger or more comprehensive universities, different schemes may be necessary for different fields.

When properly done, program evaluation can be of great benefit to a university. It informs decision-making, allocation of resources, and progress against mission.

Importantly, it makes visible to stakeholders the progress and contributions of the university, and its effective role in economic development and social good. It can also demonstrate standing and upward progress towards aspirational peers and be a key part of accreditation and national quality reviews.

To benefit from the investment of time and resources necessary for the review, it is crucial to act on the outcomes of evaluation exercises. In fact, evaluation and subsequent action are two complementary phases of a process that moves the university closer to its goals. In designing the evaluation process, the university must address the question: Who is responsible for using the evidence to bring about improvement? Some evidence will address topics related to the supporting practices of Chapter 7: the effectiveness of strategy, the functions of governance, etc. These recommendations should be fed back to the senior leadership and administration. Other evidence will pertain to the academic practices in education, research, and innovation. Improvement in these domains rests predominantly with the faculty, acting as individuals or collectively, and is discussed below.

8.2.2   Factors in the Design of Program Evaluation

The evolving view of Program Evaluation is shaped by several factors:

- A shift in emphasis from scholarly impact to societal impact, and to longer time-frames needed to observe such impact.
- Expansion of peer-review-based evaluation to include education and innovation, in addition to research.
- Development of quantitative as well as qualitative measures.
- Understanding that measurement will precipitate aligned action.

We discuss these factors below and propose evaluation approaches that address them.

It is no longer sufficient to cite the impact of universities solely on scholars and students. Stakeholders want to understand universities' broad societal impact [13]. This requires evidence of knowledge exchange across the university boundary, plus its subsequent uptake by partners and society. This takes time. With time, the potential impact grows. But longer timeframes make it harder to demonstrate causal links between the actions of the university and changes in society.

We propose a scheme that strikes a balance by dividing the evaluation into three timeframes (Fig. 8.1). The university's long-term goals are important, but difficult to measure. Therefore, we suggest intermediate measures *linked to knowledge exchange* that will indicate if the university is on a trajectory towards its long-term goals. Then, we recommend near-term outcome measures that will likely lead to the intermediate measures. The three timeframes are these:

- Near-term refers to the relatively short period in which first evidence of academic outcomes can be collected within the university and from close peers. This

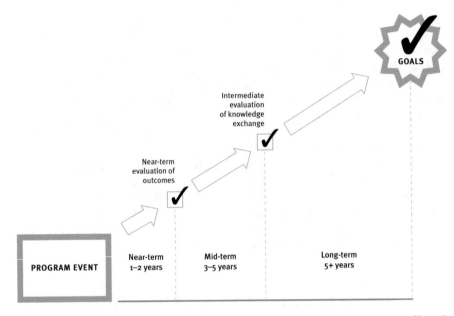

Fig. 8.1 A framework for near-term evaluation of outcomes and intermediate evaluation of knowledge exchange as indicators of that a university program will reach its long-term goals

includes evidence of educational program operations, research discoveries, and innovations. The natural cycle of this near term is probably a year or two.

- Mid-term refers to an intermediate period when the first evidence of knowledge exchange with external partners becomes available. Early impact outside of the university can more easily be correlated, but is weaker. Waiting longer produces stronger impact, but the causality can be more difficult to document. A balance might be a 3- to 5-year cycle. This timing is reasonable considering the length of degree programs, the delay until a discovery is cited, and the time required for a start-up's first commercial products.
- Long-term looks to the more distant future. It tries to identify substantial progress towards the goals of the university and its impact on society. This is likely to require more than 5 years.

In the sections below, we will further discuss these timeframes and the types of evidence that can be collected within them.

Many of the current output-based measures reflect mainly research achievement. One reason for this is that research works within a standardized peer-review process. To normalize the situation, the community needs to develop measures building on peer review for education and innovation. That peer-assessment process is discussed below.

The outcomes of universities are diffuse and complex, so they need a mix of evaluation approaches that blend quantitative and qualitative measures. The quantitative measures are easier to compile and compare, and often include both quantity and quality descriptors. For example, a quantitative measure might be some target number of articles per faculty per year (the quantity) in first quartile journals (the quality descriptor). These quantitative measures are often elaborated by qualitative indications of success. These qualitative descriptions put a human face on the quantitative data and can highlight impact and areas in need of improvement. This wider range of measures is necessary to describe the outcomes of a research university that embraces innovation.

The university needs to decide whether to evaluate and take actions based on a set of Key Performance Indicators (KPIs). In some national systems, KPIs and target values are required to provide evidence of the value of public spending [14]. Some universities have adopted them voluntarily. Such systems certainly supply countable indicators, but may produce unexpected consequences. If used, a small number of KPIs should be set at a high level to allow the leadership flexibility in evolving the university while meeting goals. The KPIs will reflect progress towards the goals of the university—if they incorporate near-term measures of academic outcomes and intermediate measures of knowledge exchange.

Preferably, a university would first define its evaluation measures based on its goals. It would then benchmark the performance of aspirational peers against those measures. For example, if a university aspires to be a small, fundamental scholarly institution, it could select a handful of recognized peers around the world and identify their performance. The university would then set its own targets informed by the peers' performance.

The use of evaluation measures will generally influence the behavior that is measured, as individuals and organizations naturally work towards meeting evaluation criteria. Those who craft evaluation schemes must be mindful of the consequences, as the influence can be positive as well as negative. Evaluation criteria should be aligned with university goals and clearly communicate them. In order to learn from peers, universities would do well to benchmark the evaluation systems of other universities.

Below, we offer suggestions to address the issues in evaluation: the shift to societal impact of longer timeframe, the inclusion of more peer-reviewed evidence, the development of qualitative and quantitative measures; and careful alignment of measures with the expanded mission.

8.2.3   Evaluation of Education

For a research university that embraces innovation, the *long-term goal* of education is talented graduates who contribute to broad social good and economic development [15]. But evidence of meeting this goal could be spread across decades [16]. Proving cause and effect would be difficult, except perhaps through biographies or narratives.

To find indicators that are likely precursors to the long-term goal, we move to the *intermediate measure* associated with knowledge exchange in education. In Chapter 3, we identify the relevant knowledge exchange mechanism: talented graduates leaving the university and taking up relevant employment. This might be in industry and enterprise, education, research, public policy, or protection of the environment. An intermediate indicator would therefore be employment rates of graduates in relevant fields. A stronger intermediate measure would include early career success. The evaluation centers on how the knowledge and skills obtained in the university have prepared the new employees, and if they have gained professional responsibility. This information is often gathered through alumni or employer surveys. The timeframe for such surveys is probably 2–5 years after graduation.

If the university has educated them well, graduates will be ready for their early career experience. The *near-term measure* would be students achieving desired learning outcomes in fundamentals and skills. In addition, they might have benefited from learning in research, emerging thought, management, leadership, and entrepreneurship.

The foundation for this evaluation is the program learning outcomes [17]. The evaluation can take the form of reviews of standardized measures of student performance (e.g., examinations) combined with rubrics for evaluating projects, student questionnaires, and interviews. Reflective memos written by the lead instructor at the end of the semester should emphasize progress towards learning outcomes and improvements to be made in the next offering. The timeframe for this evaluation is a semester or academic year. These evaluations can be tracked over time to gain a sense of students' overall progress towards learning outcomes.

For those universities that track KPIs, a notional set is shown for education programs in science and engineering in Table 8.3. A broad but representative near-term indicator of educational value is the number (or percent) of students (the quantity) who achieve a certain level of attainment in desired learning outcomes (the quality descriptor). A consistent mid-term indicator for initial success of knowledge exchange is the number (or percent) of graduates (the quantity) who find employment in jobs related to innovation (the quality descriptor). There is a reasonable expectation that the more graduates find appropriate first jobs, the more likely the university will be contributing economically.

Who is responsible for using this evidence to bring about improvement? In education, it is a shared responsibility [18]. Some aspects of education depend on the actions of individual faculty members, and some on collective action and coordination across a program. Therefore, the evaluation information must be widely shared and reviewed for implied action. In particular, program boards would identify concrete topics for improvement by joint collegial efforts.

Table 8.3 Notional key performance indicators for near-term outcomes and mid-term knowledge exchange for a science and technology program that embraces innovation

	Education	Research	Catalyzing innovation
Outcomes	Talented Students	Discoveries	Creations
Near-term outcomes (quality descriptor)	Students (developing desired learning outcomes)	Publications (in respected journals)	Patents applied for and issued (with strong claims); Partnerships with industry (funding); Publication (joint with industry)
Intermediate knowledge exchange (quality descriptor)	Graduates employed (working in innovation)	Citations (by scholars); Innovation Impact (impact study)	Patents licensed (revenue); Personnel exchanges (scope and number); Start-ups (investment)
Desirable Long-Term Goal	Social and economic impact by graduates	Follow on discoveries and impact on societal good	Partners creating successful new goods, services, and systems that address societal good

8.2.4 Evaluation of Research

For a research university that embraces innovation, the desirable *long-term research goals* are discoveries at the frontiers of knowledge that lead to even broader follow-on discoveries [19]. Discoveries also lead to creations and towards products, processes, and systems. This impact can take decades. The causality can be partially established though citations in papers and awarded patents, but it is difficult.

Mid-term research impact requires effective knowledge exchange. As discussed in Chapter 4, the evidence of successful research knowledge exchange includes recognition by other scholars in the form of citations, invitations to participate in joint projects and discussions, and personnel exchange. These items could be cataloged by bibliometrics and by an annual self-reporting of faculty activity. The transition from discoveries to innovations would best be revealed in an impact narrative. These measures would indicate scholars are on the path to long-term research impact.

The research program of a university focuses on fundamental impact, collaboration, cross-disciplinary efforts, and work with industry on major social issues. Therefore, the intermediate measures must contain not only citations by other scholars, but also evidence of impact on innovation as might be revealed by an impact narrative [20]. The case called The UK Research Excellence Framework (REF) explains the UK university research evaluation framework that relies on such impact narratives (Case 8.1).

Case 8.1 The UK Research Excellence Framework (REF)

A systematic, periodic national quality review by peers can influence university researchers and innovators to take into account the societal impact of their work.

The Research Excellence Framework (REF) is the UK system for assessing the quality of research in higher education institutions. The 2014 REF was the first initiative to factor in the impact of research outside of academia. Impact was defined as "an effect on, change or benefit to the economy, society, culture, public policy or services, health, the environment or quality of life, beyond academia." The 2014 REF is discussed here, with the caveat that, as of this writing, the 2021 cycle is in the planning stage.

In a process like the REF, three motives coincide. The outcomes of the REF are used by the government and funding councils to incentivize and reward behavior they wish to encourage. They use the results to inform allocation of grant funding for research, to provide accountability for public investment, to provide benchmarking information, and to establish reputational yardsticks. Academics gain an independent view that allows them to assess their performance against their peers in other institutions. And other stakeholders, including businesses, have quality data to inform decisions on recruitment and research collaboration.

The REF ratings were calculated by peer-assessment panels. Institutions were invited to make submissions to 36 expert sub-panels, each working under the guidance of four main panels (see Table 8.4). The panel members were appointed by the UK funding bodies and included international members. At the end of the assessment, each panel produced an overview report detailing how it operationalized the criteria, and providing general observations about the assessment and the state of research in its discipline areas. The 2014 REF involved 154 of the 164 higher education institutions across the UK submitting the work of more than 52,000 researchers, including more than 191,000 research outputs, and nearly 7000 impact case studies.

In the assessment, research *excellence* accounted for 65% of the rating and was judged in terms of the originality, significance, and rigor of the research, with reference to international quality standards. Research *impact,* as defined above, represented 20%, and was judged based on a case narrative submitted. The remaining 15% reflected research *environment* factors, such as facilities and career structure.

Table 8.4 Participants and structure of the REF evaluation process

4 main panels	4 panel chairs
	23 international members
	17 user members
36 sub-panels	36 sub-panel chairs
	1052 members

Most universities affirmed that they use the REF to internally manage research performance. Academic institutions can get a broad perspective of their strengths and weaknesses from the REF outcomes. The external scrutiny and benchmarking complement internal performance management and support strategic planning and decision-making.

The national research assessment process was initiated in the mid-1980s. It evolved to the Research Assessment Exercise which was replaced in 2014 by the REF. The significant new feature of the 2014 REF was the impact assessment, and its application to all submitted research, from physics to philosophy. This was initially an issue of some concern to academics, but in the end many discovered that their research indeed was making an impact.

Although there are short-term cosmetic tactics that can be used to maximize REF performance, universities have learned that following this route may not be harmonious with the long-term fostering of quality research and staff development. Increasingly, universities prefer to build robust, effective, and enduring solutions for improving impact [1].

Reference

1. Department for Business Energy & Industrial Strategy (2016) Building on success and learning from experience—an independent review of the research excellence framework. https://assets.publishing.service.gov.uk/government/uploads/system/uploads/attachment_data/file/541338/ind-16-9-ref-stern-review.pdf. Accessed 20 Jan 2020

The *near-term measure* of research outcomes would be publications in respected journals of new knowledge, data, theories, models, and predictions. The central test is whether research peers judge the research to be novel and likely to have impact on other scholars. The value of these research results outcomes is ensured by the peer-review process and can be documented by expert written evaluation—which is common at the time of promotion. Ordinarily, the first judgment is by bibliometric analysis.

For research universities that track KPIs, the usual indicator of near-term research outcomes is the number of journal articles per faculty member (the quantity) in journals of certain standing (the quality descriptor) as indicated in Table 8.3. A consistent mid-term indicator of research value is recognition in citations (the quantity) by peer scholars (the quality descriptor). Impact narratives, such as those used in the REF, are effective ways to judge mid-term impact of discoveries that have gone on to innovation impact.

Who uses this evidence to bring about improvement? Research is often a relatively individual process, so the individual faculty members are largely responsible for improvement. In larger research groups, senior faculty may take on this responsibility. The quality improvement mechanism that matters most is the feedback faculty authors receive from blind peer review.

8.2.5   *Evaluation of Catalyzing Innovation*

At a university that has embraced an expanded role in economic development, the *long-term* goal of catalyzing innovation is to produce creations—synthesized objects, processes, and systems that go on to become goods and services with economic impact. Much of the development occurs after knowledge is exchanged with industrial and governmental organizations. These entities do not openly report their developments, so the lineage of university creations will be difficult to establish except through impact narratives. These may need to be jointly developed.

Effective innovation knowledge exchange can be more successfully documented in the *mid-term,* as discussed in Chapter 5. As with research, knowledge exchange takes place through publications and discussions. Industrial researchers do not commonly acknowledge this type of precursor work. More easily observed and quantified are joint projects and personnel exchanges. Other instruments that give measurable impact of knowledge exchange include licenses negotiated, and revenue received on patents and other intellectual property. There is also tangible research property that is exchanged, and faculty are involved in consulting. University-linked start-ups that attract investment and produce revenues are a good intermediate measure. Increasingly, those in industry are being asked to serve as references in promotion cases, testifying to knowledge exchange impact. A university that can cite this kind of evidence is on a trajectory to innovation impact.

A *near-term* measure of catalyzing innovation is patent applications that have been filed as provisional and utility applications, and then issued. Patents can be judged by the strength and breadth of their claims. University partnerships with industry and use of industry funding can also serve near-term measures, as can the technologies, methods, and concepts produced. Publications jointly authored with industry authors are good near-term indicators, as are early stage start-ups.

To track catalyzing innovation, the near-term and intermediate measures just mentioned can be abstracted as KPIs (Table 8.3). They are more diverse than those for education and research. Near-term options include number of patents issued (the quantity) with strong claims (the quality descriptor), partnerships with industry at a level of funding and publications with industry colleagues. Mid-term measures could include patents licensed (the quantity) at a level of revenue (the quality descriptor), personnel exchanges of a certain scope and number, and start-ups that have attracted a level of investment.

Who is responsible for bringing about the innovation improvements? Here again, the answer is more diffuse than for research. In the case of publications and patents, individuals or small faculty groups act together. In other cases, like industrial partnerships, units and programs are probably involved. For licensing and agreements, knowledge transfer experts have a role. Clearly, good communications and an integrated plan are required.

8.2.6   *Program Reviews, Benchmarking, Accreditation, and Rankings*

Other activities can contribute to evaluation. These include program reviews by academic peers and other stakeholders, benchmarking of peer programs, accreditation and national evaluation, and participation in university rankings.

Most valuable is the evaluation of a program by an external group of peers and stakeholders convened by the university. This stimulates thoughtful reflection upon defined missions and goals. These reviews provide wisdom from the reviewers and identify opportunities to improve, framed in a national and international context.

In designing such reviews, the considerations include:

* Scope of review (just research, research and education, innovation, personnel, etc.)
* Size of review body and range of stakeholders represented.
* Standing vs. ad hoc nature of the group—do the same reviewers return repeatedly?
* Length and frequency of the meeting.
* Coordination or not with mandatory national reviews.
* Reporting point of the review—who receives it and acts on it.

In a conventional review system, a small group of peer academics will meet as an ad hoc group every 4 or 5 years, review a program or unit, and report to the next level of governance (e.g., a head of school for the evaluation of a department). In a stronger system, ten or more peer academics, alumni, and industrial leaders will meet biannually as a standing body to review all aspects of the program and report to the senior governance body (e.g., a board of trustees).

Another valuable form of review is peer benchmarking, a powerful way to measure comparative performance. Informal benchmarking is based on knowledge of another institution gained from the media, correspondence, visits, and faculty mobility. External thesis examiners and national quality process visits are also ways to share information. We recommend a more formal benchmarking process. It should start by identifying a small number of aspirational peer programs. A combination of informal and published data and other information, including surveys, can be used to provide comparative information. The process might include a site visit to the peer university by senior leaders, an analysis of evidence, and a report.

Nearly every university is involved in a national quality system or accreditation process to set a minimum standard for programs so that recognized degrees can be awarded. Some universities will find these minimum standards a challenge; for them, the standards are part of the valid aspirational goals. Some universities will comfortably meet the standards. New or strongly established programs may feel constrained by the standards as reflections of the past. In this case, it is the obligation of the university to engage with the quality body about changing the standards, or the creation of alternatives. The quality body must also be prepared to discuss alternatives.

A number of independent agencies produce various *university rankings* by blending data reported by the university with openly available information and with information derived from surveys. Such rankings have recently commanded the attention of universities and their sponsors. They can cause universities to improve their data-gathering capacity. They can also lead to concern that the rankings will skew behavior away from self-determined or regional goals [21, 22]. It is obvious that numerical indicators cannot capture all of the richness of the impact of complex organizations such as universities [23].

External rankings need to be used carefully as part of overall evaluation [24] and should not be the sole driver of university actions [25]. Most rankings evaluate research through its impact on research peers, and capture less completely its impact on innovation [26]. Early efforts are underway to create regional educational rankings, but these rely heavily on student opinion and inputs, rather than outcomes.

8.3   Faculty Expectation and Recognition

8.3.1   Expectations as a Means of Soft Alignment

Outcomes in education, research, and innovation knowledge exchange under the control of individual faculty can be influenced by agreeing on general expectations for the faculty and by granting appropriate recognition. This approach to soft alignment between the personal actions of faculty and goals of the university respects the valued autonomy of faculty. When faculty meet these expectations, the university runs well, and faculty should be acknowledged. When faculty exceed expectations, the university prospers, and the faculty should be celebrated. This can become an important part of a virtuous cycle of improvement.

Box 8.3 Goals of Faculty Expectation and Recognition

Universities will improve their academic programs and communicate accomplishments and institutional value to stakeholders, all while adapting to their expanded mission in innovation

By establishing expectations and recognizing accomplishments of individuals in education, research, innovation, and knowledge exchange

Better aligning the goals of the university and actions of the individual faculty.

Universities are first and foremost communities of scholars, who have broad freedom to work on questions of their own selection. There is absolutely no contradiction between this freedom and the expectation of quality. For example, if faculty members work as researchers, it makes sense that they meet expectations for research quality and quantity that are established by their community. What they work on and how they work is up to them, but the quality of the outcome is an issue of common expectation.

Universities, programs, and units can create a statement of what is expected of faculty members. The expectations provide notional guidance, not rigid formulas, that can be adapted to context. The expectations reflect the activities under control of the faculty as individuals. Expectations should guide the commitment of faculty resources, energy, and time. Expectations should be clear, traceable to employment documents, and consistent with policies. Expectations are revealed in different ways depending on the evolution and culture of the university. They can be implicit and only informally articulated, they can feature in promotion criteria, or they can be stated explicitly in a document.

The expectations will vary considerably depending on the field. In addition, in a community of faculty, different roles may differently influence expectations. There will be significant variation as a career progresses. A senior professor may be asked to do more teaching, a junior professor may be encouraged to concentrate on research, while a professor of the practice might emphasize innovation.

8.3.2 Expectations in Education, Research, and Catalyzing Innovation

The main expectation in *education* is that faculty carry out the normal teaching assignment each year, while meeting the local quality standards for instruction. These standards are most likely linked to student attainment of course learning outcomes. Faculty members might also contribute to developing curriculum plans, participating in student outreach and admissions, and serving as an academic advisor and mentor.

These contributions might reflect education in emerging thought, curriculum with embedded life and professional skills, or preparation for management, leadership, and entrepreneurship (Chapter 3). The faculty member's contribution to educational knowledge exchange will emerge through effective student learning.

The accompanying case, The Royal Academy of Engineering—Career Framework for University Teaching, presents a model for supporting and evaluating the progression of an academic career based on contribution to teaching and learning (Case 8.2). The underlying expectation is that all academics who teach should continue to strengthen the quality and impact of their teaching activities as they progress through their career.

Case 8.2 The Royal Academy of Engineering—Career Framework for University Teaching

Making teaching matter in academic careers, through a framework for demonstrating, evaluating and rewarding teaching achievement.

Launched in 2018, the Career Framework for University Teaching is an open-access resource designed to guide and support academic career progression on the basis of contributions to teaching and learning [1]. The Framework rests on the principle that all academics who teach—whether they be in an education-focused role or in a blended research/teaching role—should continue to strengthen the quality and impact of their teaching activities as they progress through their career [2].

Commissioned and funded by the Royal Academy of Engineering in the UK, the Framework is the product of 4 years of benchmarking, research, review, and piloting at universities across the world. Offering both a structured pathway for academic career progression and an evidence base on which to demonstrate and evaluate teaching achievement, the Framework provides a template that universities can adapt to their career structures and progression points. It can be used to advance teaching achievement across the academic career, including appointment, professional development, appraisal, and promotion.

For each level of teaching achievement, the Framework addresses the following three questions:

- What is the academic's sphere of impact in their teaching and learning activities?
- What promotion criteria define the academic's achievements in teaching and learning?
- What forms of evidence can be used to demonstrate the academic's teaching achievements?

The Framework is structured around four progressive levels of university teaching achievement (see Fig. 8.2). Each level can be characterized in terms of the academic's sphere of impact in teaching and learning, which expands as they progress to subsequent levels. Level 1—*the effective teacher*—represents a threshold of teaching achievement which all academics should attain. Level 2—*the skilled and collegial teacher*—takes an evidence-informed approach to their development as a teacher and promotes a collegial and collaborative educational environment across their school or discipline. At level 3, the pathway splits. Level 3A—*the institutional leader in teaching and learning*—refers to significant contributions to enhancing the educational environment at their home institution, and Level 3B—*the scholarly teacher*—refers to contributions to scholarly research, influencing educational practice within and beyond their institution. Level 4—*the national and global leader in teaching and learning*—is likely to be reserved for those progressing to full professorships solely or

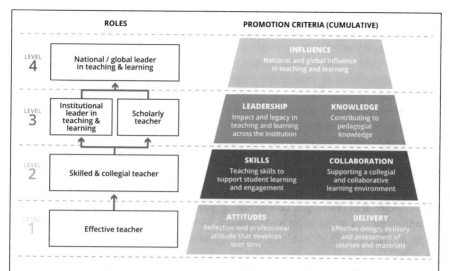

Fig. 8.2 The four levels of the Career Framework for University Teaching (left), with promotion criteria for each level (right)

predominantly on the basis of their teaching achievement. For each of these levels, the promotion criteria are conceptualized around the key capabilities that determine achievement (see Fig. 8.2).

To date, the Career Framework for University Teaching has been adopted, or adapted, by more than 30 universities worldwide, including University College London, UK; the University of New South Wales, Australia; Universiti Teknologi Malaysia, Malaysia; and Chalmers University of Technology, Sweden.

Following the launch of the Framework in April 2018, the next stage of the initiative is focused on tracking the impact of reforming university reward, recognition, and support systems with respect to teaching. In early 2019, a cross-institutional survey was launched at 15 universities worldwide that have implemented, or plan to implement, such a program of reform, to capture and track the institutional culture and status of teaching over time [3].

Contributed by Ruth Graham, R.H. Graham Consulting Inc., UK.

References

1. Graham R (2018) The career framework for university teaching: background and overview. Royal Academy of Engineering, London
2. Graham R (2016) Does teaching advance your academic career? Royal Academy of Engineering, London
3. The career framework for university teaching. https://www.teachingframework.com. Accessed 20 Jan 2020

The expectation for outcomes by an active *researcher* varies significantly from field to field. For example, some scholarly communities have an expectation of writing a book every several years; some, a lengthy journal article every year; and some, a more frequent article or refereed conference paper. The intent is that these would be key contributions to knowledge exchange, and that peers would eventually take note of these contributions.

Research expectations might extend to conducting impactful fundamental and collaborative research (Chapter 4). Faculty are generally expected to attract research funding and to supervise post-doctoral researchers and graduate research students.

In *innovation,* there is no single main expectation. Instead, there is a more diverse list. Faculty might file a patent or disclosure linked to an invention. They might become an entrepreneur and participate in a start-up. They might actually work as an innovator and develop a new product, process, design, or system. They could become involved in maturing a technology. More generally, they might be involved in an applied or industry-related project that could lead to a patent, start-up, or product.

In addition, faculty might help teach in innovation courses, interact significantly with industry, or supervise student innovation projects. These all contribute to developing an understanding of industry needs and to help exchange knowledge of university creations. The impact of these activities could be confirmed by a review involving partners in industry and enterprise (Chapter 5).

Good practice suggests using an instrument for reflection on these expectations. It could be an informal annual self-reporting that allows the faculty member to reflect and to set personal goals for the upcoming year. In some systems, the annual performance and compensation review is more formulaic. In others, the compensation is based on setting and attaining personal goals that can be linked to the expectations. Reflections on expectations can support a discussion that provides feedback by a mentor or supervisor and highlights possible recognitions.

Case 8.3 provides a view of a recently structured set of incentives and expectations at a young university—the King Abdullah University of Science and Technology (KAUST). The incentives cover the wide range of faculty activity and are aligned with university goals. They are also reflected in the promotion standards of the university.

Case 8.3 King Abdullah University of Science and Technology (KAUST)— Faculty Recognition and Appointments

By setting high expectations, providing adequate resources, and adopting a transparent and flexible promotion and rolling tenure program, a young university is able to recruit a faculty that demonstrates excellence.

KAUST aspires to be a destination for scientific and technological education and research. By inspiring discoveries that address global challenges, it strives to serve as a beacon of knowledge that bridges people and cultures for the betterment of humanity. In addition, KAUST seeks to be a catalyst for innovation, economic development, and social prosperity in Saudi Arabia and the world.

Faculty members are expected to participate in a wide array of activities as part of their professional duties. These activities are typically described as teaching, research, economic development, and service. There is an expectation that faculty invigorate graduate education in an interdisciplinary, team-based environment. They serve as hubs for engagement of private-sector partners. Faculty are expected to disseminate new knowledge, acquire additional skills, establish and maintain international connections, recruit students, consult with external entities, provide service to their disciplines, and promote the activity of the university.

To attract and retain members of the faculty, the University must aspire to the highest standards of teaching and research. It must recruit a high caliber of students, offer superb research facilities, and provide internationally competitive salaries. KAUST provides generous research funding in the form of a baseline allocation. Competitive internal proposals are also solicited, which receive external reviews, and are ranked by an international panel that advises the vice president of research.

Academic staff at KAUST are assigned to the ranks of assistant professor, associate professor, professor, and professor of practice. Faculty are appointed on rolling or fixed-term contracts. Each year, every faculty member completes an activity report that highlights his or her main achievements in all areas in which they are expected to contribute.

Assistant professors are allowed 5 years before they submit their promotion package to gain promotion to associate professor. Associate professors are allowed 6 years to submit their promotion materials to gain promotion to professor. At any time during their appointment, with the support and approval of the dean of the division, assistant and associate professors may seek promotion. Professors are usually offered 5-year rolling contracts until they turn 60 years of age, where it will turn into a 5-year fixed contract. Professors of practice are usually given fixed-term contracts, subject to renewal.

The process for promotion begins when the dean receives the candidate's promotion package and sends out for letters from leaders in the research field. Upon receiving the letters, the dean can advance or stop the case. The candidate presents a research colloquium in front of colleagues from the division, the Promotion and Appointments Committee (PAC), and University leaders. Faculty belonging to the candidate's division are then invited to vote. The next step is a discussion at the PAC which is composed of the president, the academic vice-presidents, the three divisional deans, and four professors appointed by the president; the final decision is the responsibility of the president. Among aspects that are examined for promotion are contributions to research, but also to teaching and research transfer activities.

Contributed by Yves Gnanou, Acting Vice President for Academic Affairs, King Abdullah University of Science and Technology, Thuwal, Saudi Arabia.

8.3.3   Recognition in an Adaptable University

Recognition is a powerful incentive [27]. The university can recognize the accomplishments of the faculty when good and celebrate them when excellent. The recognitions do not necessarily link directly with fulfilling expectations. Rather, the expectations create the general conditions that enable the outcomes that are recognized.

Faculty members often receive recognition beyond the direct control of the university. For faculty in science, technology, and entrepreneurship, these might include obtaining sponsorship from industry because of a faculty member's deeper understanding of industrial needs, or revenue sharing from a successfully licensed patent. Other forms of recognition which seem traditional might still reflect some aspects of innovation: the receipt of new research funding because a well-prepared proposal reflects societal impact; and election to a scholarly society which may reflect some work with industry. These all should be celebrated, especially when they link to economic development.

Forms of recognition under university control can be expanded to emphasize the development mission. Promotion tops the list, and promotion criteria should be reexamined in light of the expanded mission. Professional honors can be in the form of ad hoc receptions, or more organized competitive awards. Seed funding and provision of space for new initiatives certainly encourage exploration of new links. Universities sometimes allow supplementary compensation to be paid as rewards and incentives.

8.4   Summary

Meaningful evaluation of university programs and units is not an easy task. But considering the huge investment in universities, it is appropriate to gather and report such information; at least regarding the outcomes that are associated with the university mission, and especially about those related to economic development as highlighted in this book.

Program Evaluation deals with collecting evidence about factors that reflect the goals and outcomes of the university. This evidence should span education, research, and catalyzing innovation and should emphasize the significance of knowledge exchange and greater engagement of the university with economic development. Many universities use the information gathered for self-evaluation and improvement. This allows universities to measure progress towards goals they have set for themselves. In some nations, the national accreditation system also requires the collection and sharing of information. In all cases, evaluation helps to communicate the effectiveness and contributions of the university to its internal and external stakeholders.

Faculty Expectation and Recognition involves establishing expectations in education, research, and catalyzing innovation, including knowledge exchange within these practices. The expectations should be established by the consensus of the scholarly community and depend on local norms. The expectations should also be aligned with a complete view of university goals. Within a culture that highly values the independence and the prerogatives of the faculty, this is an approach to soft alignment between the actions of the faculty and ambitions of the university. Recognition of accomplishments, particularly those that highlight contributions towards economic development, can be strong factor in support of change.

References

1. Patil AS, Gray PJ (2009) Engineering education quality assurance: a global perspective. Springer
2. Valero A, Van Reenen J (2019) The economic impact of universities: evidence from across the globe. Econ Educ Rev 68:53–67
3. Leydesdorff L, Etzkowitz H (1997) Universities and the global knowledge economy: a triple helix of university-industry-government relations (science, technology, and the international political economy series). Thomson Learning, Boston
4. Cortese AD (2003) The critical role of higher education in creating a sustainable future. Plan High Educ 31:15–22
5. United Nations Evaluation Group (2017) Principles for stakeholder engagement by United Nations Evaluation Group. www.uneval.org › document › download. Accessed 3 Dec 2019
6. Bryson JM, Patton MQ, Bowman RA (2011) Working with evaluation stakeholders: a rationale, step-wise approach and toolkit. Eval Program Plann 34:1–12
7. The Organisation for Economic Co-operation and Development (2017) Report on benchmarking higher education system performance: conceptual framework and data. Enhancing higher education system performance. https://www.oecd.org/education/skills-beyond-school/BenchmarkingReport.pdf. Accessed 25 Nov 2019
8. Tavenas F (2004) Quality assurance: a reference system for indicators and evaluation procedures. European University Association, Brussels, Belgium
9. Astin AW, Antonio AL (2012) Assessment for excellence: the philosophy and practice of assessment and evaluation in higher education (the ACE series on higher education), 2nd edn. Rowman & Littlefield Publishers, Lanham, MD
10. Brodeur DR, Crawley E (2009) CDIO and quality assurance: using the standards for Continuous program improvement. In: Patil AS, Gray PJ (eds) Engineering education quality assurance: a global perspective. Springer, pp 211–222
11. Goddard J, Hazelkorn E, Kempton L, Vallance P (2016) The civic university: the policy and leadership challenges. Edward Elgar Pub, Cheltenham
12. Altbach PG, Reisberg L, Rumbley L (2009) Quality assurance, accountability, and qualification frameworks. UNESCO, Paris
13. UNESCO (2005) Accreditation and the global higher education market, Paris, France
14. Stensaker B, Harvey L (2010) Accountability in higher education: global perspectives on trust and power (international studies in higher education). Routledge
15. Etzkowitz H (2014) The entrepreneurial university wave: from ivory tower to global economic engine. Ind High Educ 28:223–232
16. Ahmad NH, Halim HA, Ramayah T, Rahman SA (2013) Revealing an open secret: internal challenges in creating an entrepreneurial university from the lens of the academics. Int J Concept Manage Soc Sci 1:2357–2787

17. Council For Higher Education Accreditation (2003) Statement of mutual responsibilities for student learning outcomes: accreditation. Institutions and Programs, Washington DC
18. Banta T, Palomba C (2014) Assessment essentials: planning, implementing, and improving assessment in higher education, 2nd edn. Jossey-Bass
19. Clark B (1998) Creating entrepreneurial universities: organizational pathways of transformation. Emerald Group Pub Ltd
20. Dowling DA (2005) The Dowling review of business-university research collaborations. Royal Academy of Engineering. https://www.raeng.org.uk/publications/reports/the-dowling-review-of-business-university-research. Accessed 28 Oct 2019
21. Shin JC, Toutkoushian RK (2011) The past, present, and future of university rankings. In: University rankings: theoretical basis, methodology and impacts on global higher education. Springer, Dordrecht, pp 1–16
22. Locke W, Verbik L, Richardson JTE, King R (2008) Counting what is measured or measuring what counts? League Tables and Their Impact on Higher Education Institutions in England, Report to HEFCE. Bristol, UK
23. Hazelkorn E (2011) Rankings and the reshaping of higher education: the Battle for world-class excellence. Palgrave Macmillan, New York, NY
24. van Raan AFJ (2006) Challenges in the ranking of universities. In: Sadlak J, Cai LN (eds) The world-class university and ranking: aiming beyond status. UNESCO-CEPES, Institute of Higher Education, Shanghai Jiao Tong University, China and the Cluj University Press, Bucharest, pp 81–123
25. Goglio V (2016) One size fits all? A different perspective on university rankings. J High Educ Policy Manag 38:212–226
26. Vernon MM, Balas EA, Momani S (2018) Are university rankings useful to improve research? A systematic review. PLoS One 13:1–15
27. Fung D, Gordon C (2016) Rewarding educators and education leaders in research-intensive universities. Higher Education Academy, York

Chapter 9
Alignment by Partners with the Adaptable University

9.1 Alignment by Partners

9.1.1 Reciprocal Alignment

The effective practices presented in the previous chapters are intended to influence the actions of the university, making it adaptable to evolving stakeholder needs and somewhat reshaping its internal and external functions. One such function is Engaging Stakeholders, where universities work to understand stakeholders and their needs. To be effective, the university's *partners* must take reciprocal action leading to *alignment*. Partners are the stakeholders with whom the university actively exchanges knowledge. Alignment occurs when the partners adopt practices that allow them to understand, support, and benefit from the university. Like the university, the partners must be adaptable, so that alignment can take place through give and take by both sides [1].

> **Box 9.1 Objectives of Alignment by Partners**
>
> The objective of alignment by partners is to ensure that the outcomes of the university lead to action by partners, and eventually, benefit to society. This requires that the partners understand the university's capability, support its development and outcomes, and ensure that the outcomes are successfully absorbed and employed.

Positive reinforcement will occur in the ecosystem if the university is interested in the needs of the partners, and the partners are interested in the needs of the university [2]. The university and partners will gain when they cross organizational and cultural boundaries, exchanging knowledge about their needs, capabilities, and outcomes. Such knowledge exchange will impact innovation, entrepreneurship, and eventually, economic development.

© Springer Nature Switzerland AG 2020
E. Crawley et al., *Universities as Engines of Economic Development*,
https://doi.org/10.1007/978-3-030-47549-9_9

9.1.2 The Partners in Knowledge Exchange

Historically, the university's partners have been understood to be *industry* and *government*. Together with universities, these form the "triple helix [3]." Industry, which includes new large companies, absorbs the university's talented graduates, discoveries, and creations. Today, industry is increasingly joined by *small and medium enterprises* as partners. Industry and enterprise span manufacturing, services, operational systems, and the creative sector.

We also distinguish between the "*government*," which shapes policy, regulates and provides base funds, and "*government institutions*," which are more like industry in that they fund projects, absorb university outputs, and produce services and systems that benefit the public. Government can be local (ranging from the community to the city), regional, national, or supra-national.

We add *philanthropies* to the mix of partners because they often play critical roles in new initiatives by adaptable universities. Philanthropies include major foundations such as the Gates, Wallenberg, and Tata Foundations, and the Wellcome Trust. Philanthropy also includes individual alumni who have strong attachment to their alma maters and often donate significantly.

These partners are many of the stakeholders shown in Table 7.3. The university and these partners must work together to ensure that the outcomes of the university flow to identifiable benefits for the partners. Such a flow will provide benefits to society and increase the respect and support for the university. While it is far beyond the scope of this book to examine these partners' internal functions, it is legitimate to ask how they can benefit from active engagement with a university.

9.1.3 The Practices of the Active Partners

In this chapter, we consider three *practices for the partners*, to complement the practices that operate within the university. The practices are summarized in the Table 9.1 and are discussed below for industry and enterprise, government, philanthropies, and alumni. Key approaches to alignment of universities and partners are summarized in Table 9.2.

While we have described these relationships as bilateral, there is significant benefit to multilateral endeavors. For example, universities, industry, and government can engage in making industrial policy (Case 5.2 on KAIST). Universities can spin off small enterprises that mature technologies and are subsequently acquired by industry. Universities, governments, and philanthropies can join in project investment.

Table 9.1 The practices for partners in aligning with universities

Practice name	Description of the practice > and its outcome
Understanding the university's needs and capabilities	Participating with the university in high level and engaging dialog > building a better understanding by partners of the needs, capability, culture, offerings, and strengths of the university[a]
Building the university's capacity to contribute	Helping to define educational outcomes, engaging in research and innovation projects, providing mentors, advising on policy, and contributing financially > increasing building the university's capacity to contribute
Developing the partner's capacity to absorb	Assigning to liaison roles personnel who can facilitate knowledge exchange between the university and the partner's organization > accelerating the effective absorption and employment of the university's talented graduates, discoveries, and creations[b]

[a]This complements the responsibility of the universities to understand their partners, discussed in the practice on Engaging Stakeholders in Chapter 7
[b]This complements the responsibility of universities to liaise effectively with their partners, discussed in the practice on Facilitating Dialog and Agreements in Chapter 5

Table 9.2 Approaches for partners in aligning with universities

	Partners improving their understanding of universities	Partners supporting the development of universities' capacity	Partners absorbing beneficial universities' outcomes
Alignment by industry, and small and medium enterprise	High-level dialog to acknowledge differences and seek an array of interactions with common benefit	Industry provides human, technical, and financial resources for programs that address industrial and enterprise needs	Industry absorption is managed by an engagement team that includes a senior leader, and specialists in HR, technologies, and markets
Alignment by government and community	Dialog to address tensions and develop interactions and programs of mutual benefit	Government creates supportive policy and provides core support for education, research, and innovation	Each level of government has a liaisons that help absorb university outcomes and oversee the flow to society of others
Alignment by philanthropies and alumni	Broad dialog to align the goals of philanthropies and alumni with the aspirations of the universities	Philanthropies provide targeted funding that often allows universities to nucleate new ventures and facilities	Philanthropies and alumni gain no direct benefit, but their efforts address their goals and produce impact in society

9.2 Alignment by Industry, and Small and Medium Enterprise

9.2.1 Industry and Enterprise Developing a Mutual Understanding with Universities

Industry and enterprise stand to be the greatest direct beneficiaries of the outcomes of a university, with an expanded mission in economic development [4, 5]. Industry will benefit from the talented graduates, research discoveries, and the innovation creations [6]. New enterprises spun off the university will add to the entrepreneurial ecosystem [7].

Likewise, industry and enterprise are key partners for universities [8]. Industry supports the university in many ways—through advice, financial, and policy support. Personnel exchanges into and out of industry are key to knowledge exchange (Chapter 5, Facilitating Dialog). Universities should work closely with industry and enterprise when the effort is consistent with social good and community goals [9].

However, universities and industries differ substantially in their objectives, attitudes, governance systems, incentives, and timescales of action. These differences can lead to an unnecessary mistrust [10]. Academics often perceive industry as interested in short-term outcomes that are incompatible with the academic mission. Industry may fear relying on universities for critical research or innovation results [11].

There is much to be gained by a well-conceived closer interaction; most importantly, many unnecessary fears and false perceptions fall away. Both the university and industry stand to gain here. Enhanced knowledge exchange not only benefits industry [12], it benefits universities as well [13] because industry can endorse their value to government and the public.

For this reason, we believe it is essential that universities and industry, explicitly and at a high level, acknowledge their differences, and seek a broad array of interactions for mutual benefit. Much of the framework developed in previous chapters is specifically intended to bring universities and industry and enterprise closer together. Exploring more comprehensive informal and formal modes of collaboration is critical, even though such collaboration is resource-intensive, time consuming, and rarely immediately productive. Industry will participate if its perception of long-term value gained is worth the effort. Enterprises, with shorter timescales, will want more immediate results.

9.2.2 How Industry, and Small and Medium Enterprise, Can Support the University

There is much that industry can do to assist universities. In general, industry can provide human, technical, logistical, and financial resources to enhance aspects of education, research, and innovation.

In education, industry can express its needs for the knowledge and skill of the graduates, which occurs in the practice of Integrated Curriculum. In the practices of Preparing for Innovation, enterprises can supply lecturers, mentors, and case study materials. Industry and enterprise can also provide student internships in innovation and entrepreneurship (Chapter 3).

In research, it is beneficial to the university leadership and researchers when industry articulates its long-term research needs. These needs inform scholars through the practice of Impactful Fundamental Research. In the practice of Centres of Research, Education, and Innovation, industry can play an active role in defining the scope of the research project and supporting its deployment and implementation. Industry can also supply realistic test environments, infrastructure, and simulations (Chapter 4).

In catalyzing innovation, industry and small and medium enterprise can be fully engaged.

In the practices of Maturing Discoveries and Creations, industry and enterprise can play a role in mentoring, as well as contributing knowledge of the potential market for the technology under development. They can also suggest how the proof of concept can best reduce risk and support adoption. They can supply effective project managers. In University-Based Entrepreneurial Venturing, small and medium enterprise can provide mentors and business ideas, and support the creation of new venture spinoffs of the university (Chapter 5).

Industries can effectively support universities by:

- Communicating their stake in universities, and their willingness to work deeply and broadly with universities that are responsive to industry's needs.
- Taking part in the assessment of university programs, giving objective and independent evidence.
- Willingness to participate in governance and advisory roles within the university.
- Providing financial resources to complement public funding, student fees and philanthropy.
- Championing universities' successes publicly, to local and regional government, and in formal and informal business forums.

9.2.3 How Industry Can Benefit by Effectively Absorbing the Outcomes of the University

Industry and enterprise are the prime beneficiaries of universities, hiring graduates and absorbing ideas. They will benefit even more as universities accelerate innovation and promote economic development.

The outcomes of universities—talented students, discoveries, and creations—all contribute to accelerating innovation in industry. Some examples of how industry and enterprise can additionally benefit are as follows:

- Graduate recruits will possess the knowledge and skills agreed on by the university, in part, based on input from industry. Recruits will also have had the useful learning experiences associated with Preparing for Innovation and Education in Emerging Thought (Chapter 3).
- Discoveries in university research will have greater influence on the thinking and actions of industry. This stems from research practices that emphasize impact

(Impactful Fundamental Research) and focus on large-scale problems of society (Centres of Research, Education and Innovation) (Chapter 4).

- The outcomes of catalyzing innovation: start-up companies, proof-of-concept demonstrations that lower barriers to adoption, and better defined intellectual property. Industry will also gain insights useful for technology road-mapping and forecasting (Chapter 5).

Large industry, and small and medium enterprise differ in their approaches to working with universities. Large industry can expend more resources on interacting with universities, and it can often work on longer timescales. An effective approach to operationalizing benefit to a large company is to develop an engagement team within the company whose role is to facilitate the interaction. It is often led by a member of the company's leadership team, aided by a human resource specialist who focuses on attracting graduates. Depending on the nature of the interaction, specialists in research and innovation topics, including experts on markets, may also be drawn in. This group might include a graduate of the university who plays the role of a point of contact.

Small and medium enterprise (SME) is lean and more driven by short-term considerations; this necessitates a different type of engagement. Often this engagement includes a student, faculty or alumnus of the university who is a founder of the enterprise and who takes on the role of liaison. The small scale of resources at SMEs makes relationship building harder. The sheer number of SMEs also presents challenges for the university. As a result, SME partnerships tend to be more varied. SME interactions can be organized in clusters centered around larger industrial partners. Sometimes, three-way projects involving the university, industry and an enterprise can be managed by a dedicated program manager who comes from and knows the industry. In general, informal relationships with SMEs prevail.

9.3 Alignment by Government and Community

9.3.1 Government, Community and the University Developing a Mutual Understanding

Local, regional, national and supra-national governments are pervasive stakeholders of a university. Universities help governments in their mission, and governments are essential to universities. As universities strengthen their mission of societal and economic development, they will gain more government recognition and long-term support.

Universities have always contributed to the development of society and the welfare of its citizens. Today, they are expanding that role. Figure 9.1 shows the broader range of university contribution to governments:

- In the community, a university can help to build cultural vibrancy, a sense of volunteerism by students, an entrepreneurial ecosystem and local employment.

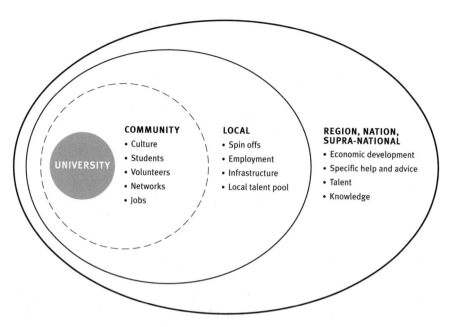

Fig. 9.1 Issues in alignment with government and community

- In local government, there are benefits from local economic development (spinoffs and jobs), contributions to urban infrastructure and an enhanced local pool of talent.
- At a higher level, the university contributes talented graduates and knowledge to economic development. Universities can respond to specific needs, for example, in climate change.

Universities create outcomes that are absorbed by government itself—government employs the talented graduates and acquires the knowledge from the university for its own purposes. Much of this is done through government institutions. These include national laboratories, mission departments and public services (e.g., NASA, the ministry of defense, national rail). These institutions help universities identify society's research and innovation needs. For example, government institutions play a key role in dealing with the environment. The relationship between universities and these government institutions is sufficiently similar to the university-industry axis that we simply reference the discussion of the previous section.

Government at all levels is critical to the university. It sets the regulatory framework within which the university operates. It establishes policy for education and expends taxpayers' money on universities, either directly by formulaic grants or indirectly by support of students and research. Local city governance has direct impact on universities as it controls zoning of land, planning permission for buildings, and public transport. The neighboring community creates the context of the university and influences staff and students' quality of life.

Whether public or private, all universities need to engage with government. In return, government should engage constructively with the universities and help them be successful.

Tensions in university-government relationships are inevitable. They stem from factors like pressure on public resources, lack of mutual understanding, and failures to fully comply with stated policy. Commonly, the concept of university autonomy produces tensions and problems. We will consider this concept, along with its counterpart, accountability, in Chapter 10.

Governments also exercise a coordinating or oversight role for public systems of universities. They can influence the degree of collaboration between universities, diversity of mission and prioritization. Coherence within government can be a challenge because education, research and innovation policies are often produced by different government organizations. Tensions can also occur between a university and its local community over housing and social accessibility issues.

All of these tensions between the university and government can be addressed. Dialog works best. Sometimes it involves government with an individual university, sometimes with a group. Genuine concern on the part of universities for societal development is a strong foundation for these discussions. The dialog will be most effective when it leads to interactions and programs that yield significant mutual benefits.

9.3.2 How Government and Community Can Support the University

Government can create an integrated and supportive policy environment, and can provide critical core resources to the university for academic and supporting practices. Government can also improve knowledge exchange and the academic practices in education, research and catalyzing innovation through its quality assurance programs.

Governments can help universities in a number of other ways, including:

- Reinforcement of an integrated government policy for the three domains of education, research and catalyzing innovation.
- Provision of a coherent funding system across government departments within an overall policy of encouraging universities to diversify their funding sources. These policies might include programs for young investigators, and for larger scale strategic efforts.
- Participation by government experts in multi-way dialog on the needs and capabilities that involve universities, industry and other relevant stakeholders. Such multi-way discussions complement the bilateral university-government dialog.
- Targeted support for innovation practices, sometimes called third-stream funding. This could include government co-funding for the practices of Maturing

Discoveries and Creations, and University-Based Entrepreneurial Venturing (Chapter 5).
- Participation in Centres of Research, Education and Innovation. These Centres bring together relevant stakeholders to produce directly implementable and impactful solutions that address issues of significant scale (Chapter 4).
- At the community and local level, government can support the university by including it in its transport, housing, and urban and land planning. Local government can use its resources and zoning powers to enable incubators and science parks.

Government can strengthen the university's impact by helping it be more functionally autonomous and robustly accountable. The supporting practices outlined in Chapter 7 build the university's strength in governance, leadership and management. These enable it to be self-governing, entrepreneurial, consciously risk-taking, and accountable. These are some of the ingredients of *productive autonomy*, as will be discussed further in Chapter 10.

Accountability for efficient and effective spending of resources is a joint responsibility. There is advantage in government and universities working together to develop well-conceived evaluation criteria. This is discussed in Chapter 8 on evaluating universities.

In all systems of modern universities, the nature and mix of university funding is critical and may be hotly contested by the faculty [14]. Public funding is important in all systems, whether it is by direct grant, indirect support of students, or by competitive award. It is especially key in science and technology. Even fully private universities take in considerable public funding. The balance between public and private funding varies by country [15]. It is beyond the scope of this book to evaluate public funding practices since they are a function of local conditions and traditions and vary enormously. However, the signals associated with the level of public funding and the ways in which it is allocated are of relevance.

The level of public funding reflects in part the priority attached to universities as drivers for development. The relative importance of the domains of education, research and innovation will be signaled by the method of allocation. The method of allocation affects the relative emphasis on different domains.

9.3.3 How Government and Community Can Benefit by Engaging and Effectively Absorbing the Outcomes of the University

Governments invest in education, research and catalyzing innovation to help meet the expectations of their citizens. Therefore, universities are stakeholders of the government. In order to strengthen public benefit, government officials should consider the following:

- Government should incorporate the contribution of universities as an integral part of overall economic policy, while recognizing that universities contribute in other spheres of society.
- Governments should collect robust information on the contribution of universities to economic development in all of its facets. Evidence gathering is essential for informed planning, for broad communication in society, and to justify continued investment.
- Many departments or agencies of government should engage with universities, beyond the usual connection through a department or ministry of education. Governmental science and economic development institutions in particular should link with universities.
- The mission departments or ministries should become more interested in universities. Links with departments of health, defense, transport, and communications are crucial.
- Local and regional governments could integrate local universities more overtly into their development strategies. They could benefit from the university's expertise, perhaps calling upon university experts to help local start-ups. Universities could be invited to local civic organizations to build consensus on local development.

The organization of governments varies widely. In general, an office of government at each level should be assigned the role of liaising with universities. This liaison can build an understanding of the universities, and formulate links so that government can absorb benefit and oversee the flow of other outcomes directly to society. To be good partners, universities should also assign government liaisons. The university president will always have a key role in dealing with government.

9.4 Alignment by Philanthropies and Alumni

9.4.1 Philanthropies and Alumni Developing a Mutual Understanding with Universities

Philanthropies, both foundations and individuals, are major forces for societal development [16, 17], including as sources of funding for universities [18]. They are well established in the US and in some European countries, and rising in importance elsewhere [19]. Some philanthropies contribute to the development of science and technology, for example the Gates Foundation. Others focus on: building university capacity, like Atlantic Philanthropies; entrepreneurship, like the Kauffmann Foundation; and education development, like the Temasek Foundation in Singapore. Philanthropies often evolve from individuals' financial donations, and these individuals are often key partners as well.

Philanthropies can play an important role for a university, strengthening its role in economic development [20]. Because they have a small set of stakeholders, they are often more willing to take risks on new initiatives.

Foundation and university staff should engage in long-ranging and broad dialog to align the foundation's goals with universities' aspirations and capabilities. It is important that universities and foundations know each other. The onus is usually on universities to make themselves aware of relevant foundations and their criteria and methods of assessment before deciding to engage.

University alumni represent a special category of philanthropists. Their engagement with their alma mater is motivated by loyalty, by a desire to see the university thrive, and by the influence of alumni networks. They are important sources of expertise, advice and funding, but only if there is a concerted engagement by both sides. Alumni want to be kept informed so they can decide how to contribute.

9.4.2 How Philanthropies Can Support the University

Foundations' essential support of the university is financial, guided by an agenda of doing good. Concentrated foundation gifts often allow universities to nucleate entire new ventures and facilities. Therefore, foundations are an important resource for programs of change. Philanthropies complement government funds, which are often more sustaining but constrained by traditional boundaries and limitations. Foundations can also contribute their considerable expertise to the development of learning goals, strategic planning, evaluation of programs and mentoring.

Alumni benefit universities in multiple ways, especially in the mentorship of students and staff interested in contributing to economic development. They can make social networks available for students, advise the leadership, and participate in governance. They can make financial gifts ranging in size from modest to blockbuster.

By definition as givers, philanthropies garner no direct benefit to themselves by their investments. Their contributions, both financial and in-kind, address their goals and potentially benefit society broadly.

9.5 Summary

The supporting practices in Chapter 7 stress the importance of university engagement with relevant external partners in advancing Knowledge Exchange. This chapter challenges the most important partners—industry and enterprise, government and philanthropies—to play their parts.

We have focused on three practices common to the partners. These could be adapted by each partner to align with the university's actions and deal with tensions. First, the cornerstone for success is partners' understanding how the university works, what it uniquely offers and what are its needs. The second practice is actively contributing to the university's capacity to better deliver on its mission and meet its goals. This is the converse of the university's actively working to meet partners' needs. The third practice is the partner developing the internal capacity to effectively absorb the university's talented graduates, discoveries and creations.

All partnerships involving different priorities, missions, timescales, and cultures can produce tension. If handled systematically and sensitively, tension can be creative rather than destructive. Conflicts can be dealt with based on a commitment to be open, honest, and action-oriented. The variety of the partners make for a stimulating and enriching experience for all.

References

1. Wright R (2008) How to get the most from university relationships. MIT Sloan Manag Rev 49:75–80
2. Owen-Smith J (2018) Research universities and the public good: discovery for an uncertain future. Stanford Business Books, Palo Alto, CA
3. Etzkowitz H (2008) The Triple Helix: university–industry–government in action. Routledge, London
4. Yusuf S, Nabeshima K (2007) How universities promote economic growth. The World Bank, Washington DC
5. Saxenian A (1996) Regional advantage: culture and competition in Silicon Valley and route 128. Harvard University Press, Cambridge, MA
6. Roessner D, Aviles CP, Feller I, Parker L (1998) How industry Benefits from NSF's engineering research Centers. Res Technol Manage 41:40–44
7. Rose D, Patterson C (2016) Research to revenue: a practical guide to university start-ups. In: Hodges TLH Jr, Hodges LH Sr (eds) Series on business, entrepreneurship, and public policy. The University of North Carolina Press, Chapel Hill, NC
8. Parker LE (1992) Industry-university collaboration in developed and developing countries. The World Bank, Washington, DC
9. Goddard J, Hazelkorn E, Kempton L, Vallance P (2016) The civic university: the policy and leadership challenges. Edward Elgar Pub, Cheltenham, UK
10. Dowling DA (2005) The dowling review of business-university research collaborations. Royal Academy of Engineering. https://www.raeng.org.uk/publications/reports/the-dowling-review-of-business-university-research. Accessed 28 Oct 2019
11. Perkmann M, Walsh K (2009) The two faces of collaboration: impacts of university-industry relations on public research. Ind Corp Chang 18:1033–1065
12. US Department of Commerce (2013) The innovative and entrepreneurial university: higher education, innovation & entrepreneurship in focus, Report by US Department of Commerce. Washington, DC
13. Bickard M, Vakili K, Teodoridis F (2019) When collaboration bridges institutions: the impact of university–industry collaboration on academic productivity. Organ Sci 30:426–445
14. St.John EP, Parsons MD (2005) Public funding of higher education: changing contexts and new rationales. John Hopkins University Press, Baltimore, MA
15. Jongbloed B, Vossensteyn H (2001) Keeping up performances: an international survey of performance-based funding in higher education. J High Educ Policy Manag 23:127–145
16. Acs ZJ (2013) Why philanthropy matters : how the wealthy give, and what it means for our economic Well-being. Princeton University Press, Princeton, NJ

17. Frumkin P (2010) The essence of strategic giving: a practical guide for donors and fundraisers. University of Chicago Press, Chicago, IL
18. Murray F (2013) Evaluating the role of science philanthropy in American Universities. In: Lerner J, Stern S (eds) Innovation policy and the economy. University of Chicago Press for the National Bureau of Economic Research, Chicago, IL, pp 23–59
19. Johnson P (2018) Global philanthropy report: perspective on the global foundation sector. Harvard Kennedy School, The Hauser Institute for Civil Society at the Center of Public Leadership. Cambridge, Massachusetts
20. Vest CM (2006) Industry, philanthropy, and universities: the roles and influences of the private sector in higher education. UC Berkeley: Center for Studies in Higher Education, Berkeley, CA

Chapter 10
Embracing Change at the Adaptable University

10.1 Making Change Happen

10.1.1 Change Is Necessary and Possible

For universities that have chosen to embrace a closer relationship with societal and economic development, change is likely to be necessary. The main change is the increase in intensity of the bidirectional sharing of knowledge at the boundary of the university—what we call knowledge exchange. There are other drivers for change. The expectations of partners in industry, enterprise, and government will intensify. The hopes of the larger set of stakeholders will evolve. The university will attract students and faculty with new interests. Programs may be evaluated by new methods and criteria.

Our assertion is that change is both necessary and possible. In Chapter 1, we discuss the evolution of universities from ancient teaching academies to today's multifaceted institutions. This transformation is enormous, but it occurred over centuries. Now, the pace of change is quickening.

A fast pace is not necessarily the ally of effective change at a university. Haste may derail the very adaptation needed. However, if the university follows an *integrated* and *systematic approach*, rapid change can be achieved. This will require playing to the natural rhythms and timescales of university life and to the intrinsic innovative capacity of individual faculty (Chapter 2).

10.1.2 The Elements Supporting Change

In the previous chapters, we formulate the elements that can help to strengthen knowledge exchange and quicken the pace of change at adaptable universities. In this chapter, we link together these elements in an approach to embracing change.

© Springer Nature Switzerland AG 2020

E. Crawley et al., *Universities as Engines of Economic Development*,
https://doi.org/10.1007/978-3-030-47549-9_10

Box 10.1 Objectives of Embracing Change

The objectives of embracing change are to help the university understand the need and possibility for change, and to link together the elements that support systematic change at an effective pace.

The elements that support change are shown in Fig. 10.1 and include:

- A set of *academic practices* that universities can use to enhance knowledge exchange, and their contribution to economic development. These practices guide education, research, and catalyzing innovation. These are presented in Chapters 3–5.
- A variety of *case studies*, interspersed throughout the practices and in Chapter 6, which suggest the general applicability of the practices around the world, and in various sizes and types of universities.
- *Supporting practices*, necessary for any well-running university, but critical for one that aspires to be more impactful (Chapter 7), a scheme of associated *evaluation and action to improve* (Chapter 8) and an approach to *alignment by active partners* in industry, government, and philanthropies (Chapter 9).

For each of the practices, we describe in the various chapters a reference situation that reflects our view of the state of an average well-run university. We also outline an aspirational situation—the state where all these practices are successfully implemented. Moving from one state to the other involves blending together the various elements in a change process (Fig. 10.1).

It will be up to individual universities to define where they stand relative to the aspirational situations we describe. This analysis will determine how much change is desired. While some universities will naturally see themselves as ahead, there is no room for complacency. Every university can be better at what it is doing. Our central message here is that change is both necessary and possible. How to make it happen is the subject of this chapter.

Fig. 10.1 The elements of an overall change process from a reference to aspirational situation

10.1.3   Change as a Generic Process Adapted for Universities

Some may wonder if a generic approach to change is possible. Certainly, the path followed by a university has to respect its culture and work within its governance. But scholars of change do identify broad features of the process that are common: identifying a need for improvement and a sense of urgency, creating a vision, building a coalition, and taking first steps that are visibly impactful [1, 2]. These common features underlie our discussion of embracing change.

In a chapter on change at universities, the reader may expect a discussion of how technology may disrupt education, how artificial intelligence may replace jobs, and how the market is shifting to lifelong learning. These are certainly contemporary shifts going on in the sector. The adaptable university can respond to these shifts and prosper. Beneath these shifts are the underlying practices that we address: planning curriculum, discovering knowledge, and innovating in systems.

We will first make the case that existing universities can change to the extent required. We discuss some useful change strategies, challenges, and methods. Next, we consider change at the higher level of a whole university system, whether in a region, state, or nation. Finally, we discuss the founding of a new university, which can be considered a special case of *changing a university system.*

10.2   Change in Established Universities

10.2.1   External Permanence and Internal Fervor

It is a common perception that universities are surrounded by a sense of permanence, that they and are slow to change, and that academics resist change [3]. The argument goes that with this resistance to change universities are in danger of becoming obsolete or being bypassed by nimble newcomers.

We do not believe that this future is a given. As outlined in Chapter 1, universities have evolved from small teaching-only communities to broad institutions meeting complex expectations from society, not the least of which is to educate a large proportion of the youth. How has this evolution been possible through times of famine and plenty, war and peace, revolution and stability? While permanent, universities have adapted to just about every challenge known to mankind.

From the outside, it is easy to understand this appearance of permanence. On the surface, little seems to change from generation to generation, except for the number of students and the construction of new buildings. Students still enter, progress, graduate, and get jobs. Professors still publish papers. How can we call this unchanging institution change-oriented and adaptable?

The answer lies beneath the surface at the level of individual faculty, where the heat of fervor is found. Instructors and professors are like little agile enterprises within an umbrella organization. They regularly seek new ways of thinking and new opportunities. They design and teach new courses. They conduct research of their own design and often attract their own funding. They respond to numerous opportunities within the academic community and in broader society. The *system* may seem static. The *faculty* are agile.

Much faculty work can be seen as pilot projects; these are often run as grassroots activities, full of tiny flickers of innovation. Some of these pilots will be sustained in a department or school as new courses or research topics. Sometimes they are taken up by the university in rethinking parts of university operations. Although often invisible to outsiders, these relentless, creative bottom-up processes constitute the powerful renewal mechanism that enables adaptation of the university to new challenges, opportunities, and societal norms.

Top-down forces are the other source of change [4]. Powerful forces can flow from the top, as when a new leader is appointed, or an external crisis arrives, or new policies or funding appear. These events occur at specific moments, and it can be crucial to use them in timely fashion to make changes, if changes are desired. (The adage "never let a good crisis go to waste" has become a commonplace in our time, and it applies here.)

Top-down changes are more visible to the public. For example, in Ireland significant research funding with stringent selection criteria was introduced in the 1990s. This became a powerful driver of internal change in universities, resulting in a dramatic transformation of the Irish higher education sector in less than 10 years.

Our conclusion is that both incremental and more radical change at universities is possible. The proposals that we are making outline a further evolution of universities, and there is no reason why it cannot happen. But we also emphasize that there is no magic bullet for becoming effective in contributing to economic development. What follows in the next two sections is a pragmatic approach to the management of the necessary change.

10.2.2 Drawing on the Strengths of University Culture

In the ideal case, the change process would be well aligned with university culture, allowing the university to be both responsive and self-confident. There is great opportunity in appealing to particular values in university culture, and *drawing on them as strengths* [5]. In Chapter 7, we discuss the evolution of culture. Here, we identify core cultural factors that support change. Building on these cultural factors, listed in Table 10.1, can pave the way for successful and sustainable change. Ignoring them often leads to failure.

Collegiality is the sense of common purpose and shared responsibility. This common purpose is reflected in consensus. In the past, collegial governance dominated universities, but this has weakened as universities have grown and fragmented into smaller communities in different units and locations. Nevertheless, the desire of faculty to be consulted on any significant change and to be heard persists. Dialog and understanding make change considerably easier, and ignoring consultation is risky.

Table 10.1 Cultural factors that support change in universities

Factor	Rationale
Collegiality and consensus	Faculty want to be involved and consulted, and participate in consensus. Similarly, students can add creativity and ground truth, and administrators can contribute wisdom
Thought leadership	Faculty value proposals that are intellectually argued and well thought out. This follows from the value we attach to critical thinking and to challenging accepted notions
Evidence	Arguments presented without sound evidence will generally be rejected. This follows from the value we attach in research to the importance of evidence in supporting or rejecting hypotheses
Benchmarking	Universities respect their peers, and faculty and leaders are normally open to learning from those who are seen as aspirational peers
Impact	Faculty genuinely desire to see the impact of their intellectual contributions, and the change process is an opportunity to reflect on and strengthen their pathway to impact
Piloting	Bottom-up processes capture the innate creative nature of faculty in education, research, and innovation, and the value and joy of experimentation
Timescale	Every practice in a university has a natural pace and timescale. Change processes can be adapted to effectively benefit from this pace

A case for change will be more compelling if it is the result of thought leadership. Universities are intellectual communities where ideas are developed, disputed, and refined. The values of critical thinking and openness to challenging accepted notions are central to student development and a core characteristic of university culture.

Robust evidence makes a case compelling. A founding tenet of research is the interplay of hypotheses and evidence. Hypotheses are accepted or rejected based on evidence. Making the case for change in a university environment faces the same test of evidence.

Universities compare or benchmark themselves against other universities considered worthy of consideration. Faculty pay great attention to what their peers were doing. The cases presented in this book are a form of benchmarking. Benchmarking can be informal, gained through correspondence, visits, and faculty mobility. Other insight comes through interactions, such as the external examiner system for degrees and the national quality process (Chapter 8). A case for change is strengthened in the eyes of faculty, students, and administrators when it is backed up by credible benchmarking against respected peers.

Faculty generally work at universities to have an *impact* on society, traditionally through scholarly contributions and education. If they perceive the change process as threatening these longstanding forms of impact, they will resist. If they believe that the proposed change will broaden or strengthen their impact, and highlight it through new schemes of recognition, they can become allies in change.

Faculty have an innate tendency to pilot and to experiment at a small scale. Such local iterative change is often more acceptable than rolling out wide-scale change.

If a pilot implementation works, it can expand, and, if not, little is lost. When a pilot works, its success needs to be recognized and institutionalized; otherwise old patterns will return. However, not all change can follow this safer bottom-up route. For instance, changes to academic or administrative structures are inherently top-down.

The timescale is a critical issue. Many of the factors in this list imply time for thought, deliberation, trial, and evaluation. This runs counter to the impatience of this online age. Natural rhythms in the life of a university, for example, those associated with the length of student programs, or faculty recruitment cycles, also influence how quickly change can be realized. Rushing the pace of change may stir resistance because it makes the process seem inappropriate, or the change poorly thought out.

Despite these cultural factors that support change, obstacles will arise. This is why we describe practices and cases. For each effective academic practice in Chapters 3–5, we identify a description, brief rationale, a set of key actions and outcomes. In addition, we present cases of successful adaptation of the practices at universities around the world. With this support and an appeal to the cultural factors, change is possible. However, it does require careful management.

10.2.3 Managing Change

We emphasize key actions for managing change in a university seeking to strengthen its impact on economic development (Table 10.2). These build on an understanding of the need for change, the fervor of the faculty, and the supportive aspects of the culture just discussed. They take into account the integrated nature of the academic practices, the literature on organizational development, and our experience.

Effective and enlightened leadership (by the university president, dean, or department head as appropriate) is essential to articulate a compelling intellectual vision for change [6]. The leader engages the academic community in robust communication and dialog, identifies a broad coalition of the interested, and engages external

Table 10.2 Managing change in universities

Action	Objectives
Thoughtful leadership and vision	To articulate a compelling intellectual vision, engage the community and stakeholders, and communicate
Chartering an effective task force	To visibly empower a task force that will listen to all and design an implementation plan for change
Allocation of adequate resources	To provide sufficient sustained resources to cover the time of the task force and other key people, and the cost of pilot studies, monitoring, and iteration
Success at the first steps of staged implementation	To plan change in phases, focusing attention and resources on earlier success, while allowing mobilization for coming stages
Regular monitoring of progress towards full adoption	To identify bottlenecks and successes, to guide further development, and support full adoption of the change

stakeholders. In a university, the academic community is especially important because faculty effort is critical to bring about change.

The leader charters the process and convenes a dedicated task force made up of relevant faculty, staff, and other experts. They develop an implementation plan that includes actions and timelines. They should consider including students and external stakeholders as members of the task force, or as advisors. The task force will need firm and constant support from the leadership, and credibility among the faculty and administrators. It needs to be visible and connected to the community, and well connected at the highest level of governance. It needs to listen carefully to concerns and input, and to distill out constructive ideas without trying to please everyone.

Adequate resourcing and funding are required. Unless the task force members and those running pilots can devote significant effort to the project, momentum will be lost. Freeing the members from other routine tasks is usually necessary. Carving out adequate funds to fund pilot projects permits progress. This is often a severe challenge: results take time to emerge, but skeptics can instantly identify why money should be better spent elsewhere. Funding must cover the full time span needed to deliver sustainable results.

We recommend planning for a staged implementation. While the sustainable gains take time to emerge, achieving intermediate successes, even if modest, will create local proofs-of-concept and keep morale and enthusiasm up. Limiting the scope of the endeavor at any given time helps to focus attention and resources. It can also make the change understandable and less threatening. Trying to implement every practice at the same time risks creating change fatigue. A nominal progression of implementation for newly founded universities is discussed below, but it can be adopted as a means of staged implementation as well.

Regular monitoring involves setting the right targets and capturing all of the nuances of the outcomes that constitute progress. This is the basis for updating the university's community, committees, and governing body. Monitoring also enables timely corrective action and revision of the plan where necessary. It creates a pathway to fully adopting the change and avoiding backsliding. The steps in monitoring progress follow closely the program evaluation practices in Chapter 8.

10.3 Change in University Systems

10.3.1 Creating Conditions for Change in the University System

Every university operates as a part of a regional, national, and supra-national system. The corresponding governments have the responsibility for developing policies that maximize the benefit to society, not only of each of the individual universities, but also of the entire system. The universities are obliged to interact within this policy to meet society's diverse educational, knowledge, and economic needs.

Industry partners and philanthropies also sometimes are engaged in stimulating change of the system.

Considerations of change in university systems build on the ideas introduced above, and in Chapter 9 on alignment by active partners. Summarizing the key relevant points from Chapter 9:

- *Universities are engines for society*, contributing to social, cultural, and economic development.
- *Universities are suppliers to* government of the outcomes absorbed by the government itself—talent and knowledge, and specific help on pressing issues.
- *Government support* is critical for their success, setting policy and regulations, providing funding, and monitoring quality.
- *Coherence in government* policy and funding should match the integration of education, research, and innovation on the part of the university.

Change in the university system builds on the autonomy and accountability of the institutions. Autonomy of universities supports creativity in identifying and responding to new needs and opportunities [7]. Evidence suggests the majority of the best performing universities are also those that show the greatest autonomy [8]. Autonomous universities are the most entrepreneurial when engaging with stakeholders and generating diverse sources of income. They engage because they feel ownership of their destiny and want to develop the capacity for leadership, planning, risk management, and implementation [9].

University accountability to stakeholders justifies the autonomy universities are given. In general, stakeholder expectations for accountability are high today, and they are rising.

University systems making a strong contribution to society often have the characteristics shown in Table 10.3.

Just as each university should develop a view of its mission, vision and outcomes, government and other stakeholders should lay out a clear mission for the whole system that transcends the simple addition of contributions from the participating universities. Faculty members, with their independence, need to understand why differentiation in one topic, or collaboration on another, will be of overall benefit or lead to efficiency. Funding and incentives should align with ambitions.

A single university with a clear sense of its mission, role, and strengths can prioritize topics and focus its energies. Working with quality on those topics makes significant impact more likely. By this logic, a particular region or nation would be best served by a system of complementary institutions, each with harmonized priorities. Yet there is always an advantage in having at least two universities independently addressing important topics to avoid monopolies.

Experience suggests that the optimal intellectual contribution of a regional or national university system does not flow from having undifferentiated universities [10]. Instead, there should be mechanisms that allow some universities to excel and collect around them the contemporary thought leaders and their students. This differentiation need not be across the board; it can exist in specially focused areas. Such differentiation must not fossilize; it must be periodically assessed.

Table 10.3 Characteristics of successful university systems

Characteristic	Rationale
Ambitions	The system must have ambitions that are more than the sum of the parts—There must be benefit to scale and synergies
Complementary missions	Each institution should focus on topics within its mission and strengths—a system of complementary institutional missions will better provide efficient coverage of all needs
Differentiation	The system must allow certain areas at some institutions to excel and gain excellence, collecting thought leaders and educating future leaders
Competition	Institutions naturally compete for faculty, students, and results. Competition avoids complacency and stimulates development
Collaboration	Universities and groups at universities often collaborate to take advantage of complementary perspectives, scholars, and capital assets

Competition is a good and necessary force. Universities tend to be competitive, especially with peers in the same jurisdiction in which they compete for resources. Scholars compete to develop new results and academic programs compete for students. Local competition is important to avoid complacency and the formation of academic cartels. It stimulates change and the taking of some level of risk.

Collaboration among universities is often warranted, as is discussed in Chapter 4. It brings together collaborators from complementary or cross-disciplinary perspectives and fields, and sometimes provides access to capital assets. Collaboration is fully compatible with competition, as each process can take place between different groups or at different times. At the university level, collaboration occurs when collaborating universities see it in their strategic interest.

10.3.2   Whole System Approach to Change

When taking a whole system approach to change of a higher education system by government or other relevant body, one must consider the network of all relevant institutions and their relationships. The network landscape can be changed by altering, merging, or adding institutions. It can also be changed by altering relationships between institutions through such instruments as collaboration, joint programs, joint ventures, and joint appointments.

Consider the task of making a university system more globally competitive. Does one invest in a small number of universities so that they are competitive in most everything they do or invest in the whole system so that each university can compete in at least one area? This question is a particular challenge for small countries, and for countries beginning to develop research and innovation intensive universities.

Addressing this dilemma, Altbach and Salmi describe the features of a world-class research university: abundant resources, concentration of talent and favorable governance [11]. They also considered three routes to establishing such a university: upgrading an existing university, merging universities, and creating a new university. The last route is discussed in the next section as a special case. They recommend that any government considering the issue should carry out a whole system analysis before taking any action.

In any effort to change institutions, redrawing the relationships between institutions has value even for the best. There are many forms of constructive relationships. Program collaboration is the easiest and occurs almost without effort. Other variants involve joint programs (more institutional in nature than just collaboration) and joint ventures (self-standing organizations "owned" by the participants). Senior faculty joint appointments are a good way to create new network benefits. It is noteworthy that even well-established and independent institutions as MIT and Harvard have found it worthwhile to collaborate on several joint ventures [12, 13].

We identify several actions that a government can take to encourage its university system to develop ambitions that are more than the sum of the parts. They are anchored in the concepts of autonomy, complementarity, and competition.

The government should increase the level of autonomy granted to the universities, so that they can develop the capacity for clear vision, planning, benchmarking, and independent implementation. Government should avoid the impulse to interfere in governance or to over-regulate as problems arise. Only if problems persist should government actively engage, and then, preferably, through assistance rather than interference. Of course, robust accountability is the companion of autonomy.

To the extent possible, governments should carefully incentivize complementary missions in the system, so that each institution feels equally respected and follows a path in which it can excel. This should encourage institutions to prioritize rather than respond to all needs. But they need also be prepared to allow some efforts to excel and become differentiated.

Competition and collaboration should be encouraged. A university should identify at least one other with which it feels competitive. Having at least two good universities in a network is valuable. Benchmarking using rich and multidimensional data gives insight about the competition. Collaboration can coexist with the stronger impulse of competition if it is incentivized.

The government should act wisely, ensuring that any action taken to change the system follows from engagement with the system, and an analysis of the impact on the system as a whole.

10.4   Starting New Universities

10.4.1   The University as an Intellectual Start-up

The objective of starting a new university is to "create a dynamic intellectual organization, an institutional actor in a country's or region's higher education [14]." It can address specific needs or be a means to change the whole university system. The main actions in founding a new university are summarized in Table 10.4. While we are oriented towards economic development, much of what we discuss below is applicable to starting any new university.

A new university is usually founded around a compelling rationale perceived by the founders:

- Governments often act based on a sense of economic, social, or cultural need or a new societal role for universities, building new capability or capacity (SUTD in Singapore, KAUST in Saudi Arabia).
- Private individuals, companies, and foundations often act with a sense that the existing universities are not serving them or society well (Pohang in Korea, Olin in the USA).
- Small private groups, who are true intellectual entrepreneurs, are driven by a vision for change (Cambridge in the UK in 1208, Station 1 in the USA in 2016).

The compelling rationale often identifies gaps to be filled by the new institution. These could include an urgency to accelerate research and innovation, a need to locate a higher education institution in an underdeveloped region, or an urgency to provide greater access to education.

The compelling rationale will contain early elements of a mission, and unique vision for the new university and its strategy. It is useful to write an initial case document for the university, articulating the compelling rationale. A first draft of the formal Mission and Strategy (Chapter 7) can wait until the first handful of faculty are onboard as their ownership will be essential if the strategy is to be effective.

There is often a visionary champion who advocates for the new university. A champion can be an influential person in government, industry, philanthropy, or education. But the champion cannot succeed alone; they must be supported by a

Table 10.4 Actions in starting a university

Action	Objective
Develop the innovative vision and case for a new university	To identify the need and the compelling rationale, articulating the gaps and urgency, and the innovative mission for the new university, described by a "case for the university"
Create the extended founding team	To catalyze the effort around an individual who will champion the university, by collecting a group of "founders," choosing a founding president, and building an initial team
Plan the developmental Process of founding the university	To define a founding process that will successfully guide the initial team through the turbulent start-up phase, apply creativity to new tasks and move established tasks to more efficient standard work
Start the university as an organization	To create the organizational, legal, financial, human resources, and cultural context and processes, including stakeholder engagement, that will support the rapid and effective development of the intellectual entity
Start the university as an intellectual entity	To begin to focus on the real educational, research, and innovation outcomes of the university, starting with the three foundational practices: Integrated Curriculum, Impactful Fundamental Research, and Maturing Discoveries and Creations

variety of other individuals and organizations. Some representative set of these will formally or informally become "founders." The champion and founders rally support, create the legal entity of the university and its governing board, contribute to the vision, and recruit the founding president.

The founding president then takes on leadership responsibility, building on the compelling rationale and articulating a mission and vision. The definition of mission and choice of the founding president are closely linked—the vision will guide the choice of a leader, and the president will refine the mission. The founding President helps to establish early credibility for the university.

The founding president creates the first team of seven to ten individuals who will begin to build the university. Collectively, this team must have an understanding of social, political, and cultural context of the university, and know how to respect and work within them. They can support the president by providing organizational skills. The individuals on this first team will heavily influence the culture of the university (Chapter 7).

Starting a university is a process not unlike starting other new enterprises. On day 1, there is almost nothing but a vision, and perhaps a very small team. Over the next few years, there will follow a turbulent and iterative process of defining mission and programs, attracting faculty, staff and students, and securing resources. The organization needs a *founding process* that will guide the team successfully through these actions, naming steps and those responsible. However, the informality and novelty of this process will not be to the liking of many academics who have spent their careers in relatively placid and established universities. Therefore, it is worth recruiting people who enjoy start-ups of various kinds. Starting a university attracts a type of person who is an academic entrepreneur and who often seeks more opportunity than is available in a traditional university.

Founding a university is a time of *creativity*. The founding team will need to create all processes as needed; there are no regulations on day 1. These are built on benchmarks from other universities and on the experience of the founding team. Ultimately, you have to put the pieces together to create new solutions for the new goals. If you want certain new outcomes, you have to establish new conditions that can create them.

A new university is a new organization. Its *development process* can be styled after those processes used to create other organizations. The university could start with years of planning and with well-drafted plans. Or it could use an *agile process* with deep understanding of stakeholder needs, frequent iteration, and focus on launching the program as quickly as possible [15–17]. We recommend a mixed approach. Early participants can work collaboratively in teams around tasks. Even if plans are somewhat ambiguous, pilot implementations and engagement with stakeholders will clarify the process. After several iterations, the work can become better understood. Policies and formalized processes can be implemented that support consistent and efficient operation [18].

10.4.2 Starting the Organization and Supporting Practices

Getting the organization of the university going is the first step to fulfillment of its vision. These steps map to the six supporting practices of Chapter 7. The challenge is to create the shell of an organization that will be responsive to, and supportive of, the academic activities, even before the first professor or student arrives. The practices of Mission and Strategic Planning are discussed above. We discuss the five other practices below.

The university is more likely to create the outcomes of importance to external stakeholders and gain their support by moving quickly to engage them (Chapter 7). The founding team should identify the key stakeholders, get to know them, listen carefully, and understand what the success of the university means to them. The university should reflect on these needs in formulating goals. The university also needs to inform them of its initial plans and create credibility.

There are three specific issues to address. One is the dramatically different timescales of expectations of the political stakeholder and the university, respectively. Six months can be a long time in politics, and just an instant at a university. Early meetings with stakeholders should stress the real timeframe for results and identify early wins that will allow the university to survive inevitable criticism for being slow to deliver.

The second issue is engagement with "competitors," which are those institutions that think they should have received the support now being directed to the new university. It is preferable to engage these institutions early, and to try to identify how they can benefit from the new university.

Some new universities secure one or more established university partners to speed the new university's development. University partners can help to hire faculty, jumpstart programs, and design facilities.

In Chapter 7, we discuss evolving the culture of a university so that the community will embrace the expanded mission in economic development. In this chapter, we offer some observations about creating the initial culture of such a university. Culture is a powerful force that can be shaped, and that is largely set by the first handful of community members [19]. It is shaped by the founding president, both in the president's personal values and in choice of the founding team. It is important to have overt discussions about the desirable values for the university and how to establish them.

There are three cultures that evolve: support, faculty, and students. The culture of the support community (finance, human resources, etc.) is set early by the first administrative people and should focus on service. The faculty culture is set by the first group of academics. To the extent possible, these should be hand-picked not only for their scholarly and innovation accomplishments, but for their commitment to knowledge exchange and economic development.

The student culture will be set by the first pilot group of students, who are banking their future on the promises of the new university. They will be risk takers and early adopters and will be attracted by the mission and opportunity. If the initial students are treated as members of the founding team, they will contribute greatly to the spirit and excellence of the university.

Recruiting top-grade academics to the first ten or so appointments is probably the most important determinant of the new university's mid-term success. They will advocate for the university and give it credibility. They will start the programs and help attract the next group of faculty, including junior faculty.

The most powerful tool to attract these faculty leaders is the vision. Some well-established faculty genuinely want a new adventure and an environment, where innovation and impact are valued. Others look forward to the opportunity to be a founder. Others seek access to strong colleagues and students, new facilities, and financial support. If the new university has a strong academic partner, it can play a role in recruiting. If there is a national academic diaspora, it can be a source of faculty.

This process has a start-up problem: the first faculty want to know the colleagues they will work with, but these colleagues have yet to be hired. One approach is to create a group of highly capable and like-minded "founding faculty fellows": senior professors who have part-time appointments or spend a sabbatical at the new university. These fellows play the role of the permanent faculty for the first few years, and some will decide to take permanent positions.

Other important early hires are the first professors of the practice—distinguished practitioners recruited to the university—who will aid with the innovation agenda. Their early presence will impact the growing culture and will help establish innovation as a key role of the university (Chapter 7).

In time, governance will evolve and be formalized. Consider starting with informal shared governance, with the key academic and administrative leaders working together. Over time, a system of shared governance will evolve.

It is important that this leadership has control over internal matters: culture, calendar, curriculum, research, innovation, fiscal, and other policies. Funding should be negotiated with a long-term view; the project timeline will stretch beyond the career lifetime of the initial champion.

Academic buildings and facilities are developed based largely on local conditions. Some build the university facilities first in order to attract faculty, some build them in parallel with the initial start-up, and some use interim facilities while waiting for the program to evolve. We recommend creating a design based on benchmarks of aspirational peers, and the creation of generalized flexible space and facilities.

10.4.3 Starting the Academic Practices

Shortly after the organization is established, it is time to begin starting the eleven academic practices. This is simply too much to undertake at once, so we recommend starting first with the three foundational practices: Integrated Curriculum, Impactful Fundamental Research, and Maturing Discoveries and Creations (Fig. 10.2).

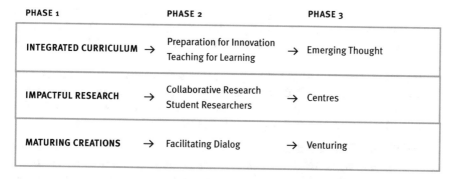

PHASE 1	PHASE 2	PHASE 3
INTEGRATED CURRICULUM →	Preparation for Innovation Teaching for Learning →	Emerging Thought
IMPACTFUL RESEARCH →	Collaborative Research Student Researchers →	Centres
MATURING CREATIONS →	Facilitating Dialog →	Venturing

Fig. 10.2 Phased start-up of the academic practices at a new university

These three practices support planning and can be done without significant antecedents. They form the foundation for the remaining eight practices. We suggest below that these eight academic practices be loosely divided into a second and third phase, with the third representing the most integrative and advanced phase. This sequencing of the three phases can also be used in the adoption or refinement of practices at existing universities. Note how all three practices build on the supporting practice Engaging Stakeholders, discussed above and in Chapter 7.

Fundamental Impactful Research can be performed individually by all faculty; it is therefore a good place to start. Recall from Chapter 4 that in this practice faculty pursue fundamental new discoveries, producing research outcomes with an impact both on scholarship and society.

The steps to implementing this foundational practice begin with:

- Commitment by the faculty member to the importance of this activity. The commitment can form part of the hiring process, and it can be reinforced in early faculty discussions and seminars.
- Reflection by the faculty on societal or industrial needs, which initially can be learned through the practice of Engaging Stakeholders. Pressing needs are often identified in policy and framing documents, created by government or organizations such as the UN and OECD.
- Prioritization of research topics that can be carried out early, before sophisticated laboratories are in place, and then the phasing-in of more experimentally intensive topics with time.

Remember from Chapter 3 that the practice of Integrated Curriculum provides faculty to collectively implement an integrated curriculum, including courses, projects, and co-curricular experiences. The end result is that students learn the disciplinary fundamentals together with the skills and approaches needed for professional practice and innovation. The main initial steps in starting this practice are:

- Developing an understanding of what student knowledge, skills and attitudes are most desired for professional practice and innovation. Again, investigating employers' needs can be part of the supporting practice of Stakeholder Engagement.

- Identifying and benchmarking the educational activities of aspirational peers, consulting with groups that develop educational approaches and standards. Like research, educational approaches should stand on the shoulders of other successes.
- Beginning the real design work by planning a framework for the degree programs (structure, content, degree of flexibility, etc.). Accreditation issues need to be addressed. As the first faculty are appointed, and working with Faculty fellows, planning of the first few programs can be drafted.

Recall from Chapter 5 that the practice of Maturing Discoveries and Creations centers on university researchers progressively making discoveries, creations, inventions, and proof-of-concept demonstrations with the aim of maturing creations. This supports their adoption by industry and in entrepreneurial ventures. The important steps to starting this practice include:

- Working with faculty to create an understanding of the need for and practice of maturing. This can be part of the recruitment process and the subject of seminars.
- Scanning the more fundamental research work of faculty members and identifying candidates for consolidation in the form of a prototype or invention.
- Using the stakeholder network, start to identify the commercial impact of such creations. It is also useful to ask what other universities the industry likes to work with, and why.

With some initial success, and as more of the permanent faculty arrive, it becomes time to move to the *second phase* of practices, which build on the foundational ones. These are achievable in several years and lead to the value outcomes of the university. We would designate these second phase practices as:

- Teaching for Learning and Preparing for Innovation (Chapter 3).
- Collaboration Within and Across Disciplines and Undergraduate and Postgraduate Student Researchers (Chapter 4).
- Facilitating Dialog and Agreements (Chapter 5).

The *third phase* activities move the university towards broad impact and excellence. They depend on the first tangible outcomes of the university, talented students, discoveries, and creations. They include:

- Education in Emerging Thought (Chapter 3).
- Centres of Research, Education, and Innovation (Chapter 4).
- University-Based Entrepreneurship (Chapter 5).

10.5   Summary

We argue throughout this book that strengthening Knowledge Exchange between universities and society enhances economic development while preserving traditional strengths and values. We have outlined a set of academic practices in education, research, and catalyzing innovation which can contribute to Knowledge Exchange. Fully implementing these good practices will require considerable effort for most universities. It should also be cause for reflection by those responsible for university systems.

This chapter is designed to stimulate an approach to adaptation and change that is sensitive to university culture and to offer reassurance that change is possible. We consider three situations: established universities, start-up universities, and university systems. Established universities comprise a wide spectrum: from specialist to comprehensive, public to private, small to large, collegial to managerial. While details may differ, the general principles we outline apply to all. Start-up universities are a special case that requires a compelling rationale, dedicated founders, and creativity. In the case of university systems where governments are inevitably involved, change must take into account the needs of society, and issues of differentiation, competition, and collaboration, as well as autonomy and accountability.

Our treatment of change is not designed to be exhaustive. There exists a broad range of literature on the management of organizational change. We have based our discussion on the scholarship on organizational change in general, while drawing from our personal and practical experiences to adapt it to change at the university.

References

1. Kotter JP (1995) Leading change: why transformation efforts fail. Harv Bus Rev 73:59–67
2. Beckhard R, Harris RT (1977) Organizational transitions: managing complex change. Addison-Wesley, Reading, MA
3. Morrill RL (2007) Strategic leadership: integrating strategy and leadership in colleges and universities, ACE/Praege. Rowman, Littlefield Publishers & American Council on Education
4. Beer M, Nohria NN (2000) Breaking the code of change. Harvard Business Review Press, Cambridge, MA
5. Musselin C (2006) Are universities specific organizations? In: Krücken G, Kosmützky A, Torka M (eds) Towards a multiversity? Universities between global trends and National Traditions. Transcript Verlag, pp 63–84
6. Crow M, Dabars W (2015) Designing the new American University. Johns Hopkins University Press, Baltimore, MA
7. Pink DH (2009) Drive: the surprising truth about what motivates us. Riverhead Books, New York, NY
8. Weber L (2010) The next decade, a challenge for technological and societal innovations. In: Weber LE, Duderstandt JJ (eds) University research for innovation. Econnomica Ltd Glion Colloquium, London, pp 37–49
9. Miller RK (2011) Some challenges of creating an entirely new academic institution, Organizational learning contracts. Oxford University Press, pp 121–138
10. Deutsche Forschungsgemeinschaft (2005) German universities excellence initiative. Deutsche Forschungsgemeinschaft
11. Altbach PG (2011) The past, present, and future of the research university. In: PG A, Salmi J (eds) The road to academic excellence. World Bank Publications, pp 11–32
12. Harvard-MIT (1970) The Harvard-MIT Program in Health Sciences and Technology (HST). https://hst.mit.edu/about. Accessed 1 Oct 2020

13. Collis DJ, Shaffer M, Hartman A (2014) edX: strategies for higher education. Harvard Business School Case
14. Altbach PG, Reisberg L, Salmi J, Froumin I (2018) Accelerated universities (global perspectives on higher education). Brill Sense
15. Denning S (2018) The age of agile: how smart companies are transforming the way work gets done. Amacon
16. Thompson K (2019) Solutions for agile governance in the enterprise (SAGE): agile project, program, and portfolio management for development of hardware and software products. Sophont Press
17. Eppinger S (2019) Ten agile ideas worth sharing. In: MIT SDM systems thinking webinar series. http://sdm.mit.edu/webinar-steven-d-eppinger-10-agile-ideas-worth-sharing/. Accessed 6 Nov 2019
18. Reppening NP, Kieffer D, Reppening J (2018) A new approach to designing work. MIT Sloan Manag Rev 59:29–38
19. Schein EH, Schein PH (2019) The corporate culture survival guide, 3rd edn. Wiley, Hoboken, NJ

About the Authors

Edward Crawley is the Ford Professor of Engineering, and a Professor of Aeronautics and Astronautics at MIT. He has served as the founding President of the Skolkovo Institute of Science and Technology (Skoltech) in Moscow, the founding Director of the MIT Gordon Engineering Leadership Program, the co-founder of the International CDIO Initiative, the Director of the Cambridge (UK)—MIT Institute, and the Head of the Department of Aeronautics and Astronautics at MIT.

Crawley's research has focused on the architecture, design, decision support, and optimization in complex technical systems subject to economic and stakeholder constraints. His book *System Architecture: Strategy and Product Development for Complex Systems* was published by Pearson (2016). His work on innovation includes helping to found six technology-based startups.

Crawley is a Fellow of the AIAA, the Royal Aeronautical Society (UK), and the International Academy of Astronautics. He is a member of the US National Academy of Engineering and is a foreign member of national academies of Sweden, the UK, China, and Russia.

He received an SB (1976) and an SM (1978) in aeronautics and astronautics and an ScD (1981) in aerospace structures, all from MIT, and has been awarded two honorary doctorates.

© Springer Nature Switzerland AG 2020
E. Crawley et al., *Universities as Engines of Economic Development*,
https://doi.org/10.1007/978-3-030-47549-9

John Hegarty served as President/Provost (2001–11) at Trinity College Dublin and, previously, as Dean of Research, Head of the Physics Department and Professor of Laser Physics. He spent nine years in the USA at the University of Wisconsin—Madison and at AT&T Bell Labs.

As President, he led a significant expansion of Trinity College's research and innovation mission, including engagement between disciplines and with stakeholders, as part of a wider change and modernization program.

Hegarty's field of research was photonics, resulting in more than 100 publications in peer-reviewed journals, three patents, and one spin-off company. He co-founded Optronics Ireland (1999), a national research program in photonics involving strong cross-disciplinary collaboration among five universities.

He co-founded Innovation Advisory Partners in 2012 to provide advisory services on research and innovation and acted as Senior Advisor to the President of the startup university, Skoltech. He is a member of several boards including those of the Irish Times, the Hugh Lane Gallery, and the Fulbright Commission in Ireland.

Hegarty received a BSc (1969) and PhD (1975) from the National University of Ireland and an ScD (2001) from Trinity College Dublin. He has three honorary doctorates. He is a Fellow of the Royal Irish Academy and the Institute of Physics, and a Fellow Emeritus at Trinity College Dublin.

Kristina Edström is an Associate Professor in Engineering Education Development at KTH Royal Institute of Technology. Since 1997, she has participated in and led educational development on a national and international level. Her focus is on enhancement of courses and curricula and of faculty teaching competence.

Edström has been active since 2001 in the CDIO Initiative, an international engineering education collaboration. She serves on the CDIO Council and leads the research track in the CDIO Annual International Conference. During 2012–2013, she was Director of Educational Development at the Skolkovo Institute of Science and Technology.

Since 2018, Edström is the Editor-in-Chief of the European Journal of Engineering Education, published by the European Society for Engineering Education. Her research takes a critical perspective on the why, what, and how of engineering education development. She investigated tensions between the academic and professional aspects of engineering education. One particular focus is on institutional conditions, such as faculty recruitment and promotion practices.

Edström has an MSc in Engineering from Chalmers and a PhD in Technology and Learning from KTH. Edström was awarded the KTH Prize for Outstanding Achievements in Education in 2004 and elected lifetime honorary member of the KTH Student Union in 2009. She has presented more than one-hundred invited talks at conferences and universities worldwide.

Juan Cristobal Garcia Sanchez has led more than one-hundred innovation projects in Mexico, France, China, and the USA for the education, high-tech, and consumer sectors. He has directed comprehensive innovation programs for six universities, with responsibilities ranging from opening new campuses and boosting learning programs to managing turnarounds that transform culture, multiply enrollment, and thrive financially.

Garcia Sanchez has worked with eight industry clusters to promote thousands of small and midsize enterprises. For the regional Secretariat of Economy, he designed a communication campaign to promote the Mexican "Silicon Valley." He led the innovation strategy for the first technology park in Mexico and for the first software center in western Mexico. He is a serial entrepreneur and co-founder of a think tank for sustainable innovation.

He is the author and editor of 20 books, some of which were government co-sponsored to support economic and cultural development. He is a recipient of the HarperCollins Award, the Grossman Award, and the MIT-SDM Award for Leadership, Innovation, and Systems Thinking.

Garcia Sanchez is a doctoral candidate at the University of Pennsylvania. He received an MSc in Engineering and Management from MIT, and a multidisciplinary MSc from New York University.

Glossary

Academic practices Patterns of human behavior in education, research, and catalyzing innovation.

Catalyzing innovation Activities of the university around synthesis, which lead to creations that have never existed in the past.

Creations Outcomes of catalyzing innovation: concepts, prototypes, inventions, processes, and know-how.

Discoveries New knowledge, facts, data, models, and theories that are outcomes of research.

Economic development Improves the prosperity, economic opportunity, and well-being of individuals, organizations, and society.

Education The process of guiding students to acquire disciplinary fundamentals, essential skills, approaches, and judgement.

Effective practices Effective because they produce outcomes that support knowledge exchange.

Engagement The development of society occupies the attention of the university, and the university is involved in and contributes to development.

Entrepreneurship The process of venture formation concurrently with the development of a first product, contributing to economic development.

Innovation The translation of lower-value resources into products, contributing to economic development.

Knowledge exchange A multidirectional flow of people, capacities, and information among relevant participants.

Outcomes The results in the domains of education, research, and catalyzing innovation that are the subject of knowledge exchange.

Partners Those stakeholders who the university engages in knowledge exchange: industry, enterprise, government organizations, cultural institutions, nonprofits, and other higher education institutions.

Products Tangible goods, services, and systems, created by industry, enterprise, or government organizations.

© Springer Nature Switzerland AG 2020
E. Crawley et al., *Universities as Engines of Economic Development*,
https://doi.org/10.1007/978-3-030-47549-9

Research The process of discovery at the frontier of knowledge, and the quest for an increased understanding of aspects of our world that were previously unknown or unexplained.

Societal development Includes economic, social, and cultural development.

Stakeholders All those who have a stake in the university, including faculty, students, the community, the public, and other partners.

Supporting practices Patterns of human behavior that support the university's efficient and effective operation.

Sustainable development Promotes prosperity and economic opportunity, greater social well-being, and protection of the environment, without compromising the ability of future generations to meet their needs.

Talented graduates Knowledgeable and skilled university graduates who are the outcomes of education.

Index

A
Academic domains, 27
Academic facilities
   catalyzing innovation, 232–234
   community, 229, 231
   education, 229, 231
   expanding role in innovation, 228, 229
   goals, 228
   research, 232–234
   spaces, 229
Academic institutions, 5
   economic development, 13, 14
Academic leadership, 218
Academic position, 78
Academic practices, 2
   decomposition, 177
   deploying, 178, 179
   university, 177
Academic practices of education
   integration, 49
   knowledge exchange, 48
   reference and aspirational situation, 49, 50
   research and catalyzing innovation, 49
Academic programs, 118
Accelerating innovation
   and entrepreneurship, 12
Acceleration phase, 197
Accelerators, 14
Active learning, 67, 86
   active engagement, 68
   belief in one's capability, 71
   experiential learning, 68, 71
   "flipped classroom" methods, 68
   PBL, 71–73
   self-efficacy, 71
Actual innovation, 141

Adaptability of universities
   economic, social and cultural
     development, 7
   expectations of government officials and
     citizens, 7–8
   governments, 6, 7
   and resilience, 5–6
Adaptable organization
   faculty and staff (*see* Facilities)
   governance (*see* Governance)
   mission and strategic planning (*see* Mission)
   stakeholders (*see* Stakeholders)
   supporting practices (*see* Supporting
     practices)
Adaptable universities
   evaluation (*see* Program evaluation)
   impact, 239
   recognition, 258
   stakeholders, 239
Alignment by government and community
   autonomy, 268
   collaboration, 268
   government institutions, 267
   regulatory framework, 267
   societal and economic development, 266
   tensions, 268
   university contribution, 266, 267
   university-government relationships, 268
   university support
     benefits, 270
     policy environment, 268
     practices, 269
     productive autonomy, 269
     public funding, 269
     resource spending, 269
     stakeholders, 269

© Springer Nature Switzerland AG 2020
E. Crawley et al., *Universities as Engines of Economic Development*,
https://doi.org/10.1007/978-3-030-47549-9

Printed in the United States
by Baker & Taylor Publisher Services